MATH TOOLKIT
for

INTRODUCTORY CHEMISTRY

Diane Vukovich • Timothy Smith

PRENTICE HALL, Upper Saddle River, NJ 07458

Associate Editor: Mary Hornby
Special Projects Manager: Barbara A. Murray
Production Editor: Mark Marsden
Supplement Cover Manager: Paul Gourhan
Supplement Cover Designer: PM Workshop Inc.
Manufacturing Buyer: Ben Smith

A Pearson Education Company
Upper Saddle River, NJ 07458

Printed in the United States of America

10 9 8 7 6 5 4 3

ISBN 0-13-010060-9

Prentice-Hall International (UK) Limited, London
Prentice-Hall of Australia Pty. Limited, Sydney
Prentice-Hall Canada Inc., Toronto
Prentice-Hall Hispanoamericana, S.A., Mexico
Prentice-Hall of India Private Limited, New Delhi
Prentice-Hall of Japan, Inc., Tokyo
Pearson Education Asia Pte. Ltd., Singapore
Editoria Prentice-Hall do Brasil, Ltda., Rio De Janeiro

CHAPTER 1

Taming Chemistry

There is a fable about a man who encountered a lion in the forest. At first he was terrified and started to run. But then he saw that the lion was limping with a large thorn in his paw. He slowly approached the lion and carefully pulled out the thorn. The lion was grateful. Years later the man was taken prisoner by the Romans and condemned to be thrown to the lions. He was led into the Colosseum and the lion, released from his cage, bounded out ferociously. But when the lion saw the man, he stopped, and instead of attacking him, he licked him in the face. The lion was the one he had befriended years before. The lion's behavior was so unexpected that both the man and the lion were set free.

What does this little story have to do with chemistry? Well, to many people chemistry may look like a beast that will devour them or at least block their way to where they want to go. But just like the lion, with a little care, you can tame chemistry, and maybe years later chemistry will come to be your friend. We hope this book will help you tame chemistry.

You Are Smart Enough to Study Chemistry

Many students believe that someone has to be extraordinarily intelligent to succeed in chemistry. And you may worry that you are not smart enough. Don't be afraid. Many normal people have successfully studied chemistry. You *do* need to have a good grasp of basic mathematics and you *do* need to be organized in your study habits. This book will help you with both of these.

Many students think that some people are "chemistry types," and they are not one of them. They label themselves or let others label them and then they begin to believe the label. Although some may love chemistry so much that they will go on to become chemists, and you might label them as "chemistry types," the chemistry in your class (and even beyond!) is well within the grasp of people with normal intelligence.

Chemistry is Relevant to Your Life

Many students believe that the chemistry class they are required to take is just a hurdle they must jump to get to their major. They believe that chemistry is only for chemists and irrelevant to their lives. But what we hope you will begin to realize is that your whole life is chemistry. Whether it's the food you eat, the medications you take, the adhesive on the back of your stamps or the ink and paper of this book, everything around you involves chemistry.

When you know more chemistry, you will be better informed to make medical, health and nutrition decisions in your own life. We are all confronted with environmental crises such as acid rain and the depletion of the ozone layer. We as citizens need to make decisions about these issues and without at least a basic knowledge of chemistry, making intelligent decisions about such matters is almost impossible.

Chemistry Plays a Big Part in Your Career

Students don't realize when they are studying chemistry what a big part it will play in their profession. One of our students, a registered nurse who came back to school to get her bachelor of science in nursing, made a list of chemistry related activities she encountered in one of her weeks on the job:

Your Chemistry Class is Not a Math Class

Many students think that a chemistry class is just another type of math class. There *is* mathematics in chemistry, but chemistry is primarily descriptive. Chemistry is a collection of observations of our everyday world along with explanations of why they occur. When you note that a loaf of bread does not come out right if you add too much or too little yeast or sugar, you are doing chemistry. Chemistry has much more in common with cooking than with mathematics. In fact cooking is chemistry.

There is not a great deal of algebra in your class and there are fewer than a dozen equations. This book will work with you on the mathematics you do need.

Your Chemistry Class Will Give You a Chance to Learn about Yourself

Your chemistry class will not only begin teaching you about the chemical processes that occur in your body—which is certainly learning about yourself—but will also teach you something about your own psychology. How do you study for tests? How do you learn to relax when taking tests? By watching and learning about these things now, you will learn something more valuable than the chemistry. You will gain valuable self-knowledge that will help you in your advanced classes and the Board tests.

Don't be Affected by Others' Negative Attitudes Toward Chemistry

Has anyone said to you when he or she found out that you are taking chemistry, "You are a glutton for punishment," or "You're in for it now"? Many people have preconceived notions about chemistry. They might have had a bad experience in high school. Or they might just be repeating what they have heard other people say. Don't be swayed by others' negative attitudes. Form your own judgment from your own experience.

In your class, surround yourself with positive people. Avoid negative people who constantly complain about the lectures, the textbook, and the homework. These students often blame others for their lack of success and refuse to take responsibility for their own destiny. They don't like to feel alone, so they will find other students to convince to feel the same way. They need confirmation for their negativity. Don't fall into this trap! Find the students in the class who have the qualities that you want to have—the students who will help you and inspire you to do your best.

Approach Chemistry With Enthusiasm

Instead of seeing your class as a burden or a chore, look at it as an exciting opportunity. Approach chemistry with enthusiasm. If at first you don't feel enthusiastic—pretend! It often happens that if we act positive and motivated, our mind starts to believe it. Tell yourself that you're really excited and eager to go to lecture or to read your textbook. Tell yourself, "I'm bright and intelligent and can learn anything I choose to learn."

Be Curious

Be curious. A chemistry class is a great opportunity to begin to learn about little things in life you may have wondered about: Why do we sweat? What's in an aspirin tablet? Why do salt crystals form little cubes? What are the ingredients in shampoo? And even, why is the sky blue? Again, if you're not curious—fake it. Just think—after finishing your class you'll be able interpret the ingredients lists on labels.

Remember: Some Confusion is Natural

Everyone is always a little confused and disoriented when confronted with new ideas and concepts. If you know students in an advanced math or chemistry class, it may not seem like it, but they struggled in their beginning courses and continue to struggle as they progress. They were probably as bewildered as you sometimes feel.

Your Success in Chemistry is Directly Proportional to Your Feeling of Self-worth

When your feeling of self-worth increases, your success in chemistry and other courses increases. This is actually a relationship similar to ones you will examine later in your class (and which we will look at in Chapter 7). When one goes up, the other naturally goes up. A high feeling of self-worth means that you believe you can succeed and not only pass your chemistry class and go on to your program and graduate, but also make a difference in your profession, win the respect of your colleagues, and truly help the people you serve.

When you are doing homework problems and studying for tests, remember what your long-term goals are. Sometimes it's easy to forget them in the everyday grind. Keep reminding yourself. And remind yourself that you are an important person with important things to do, and your chemistry class is one step on the way.

Study Aid 1 Good Study Habits

Good study habits are not among our "basic instincts." Most of us have to work long and hard to develop good study habits. Here are some concrete things you can do to promote your success as a student.

- Set a long-term goal, that is, a career goal. Goals keep us going when the going gets rough.
- Set short-term goals, that is, daily or weekly goals. Short term goals motivate us and keep us on target.
- Make a realistic daily, weekly, and term-long schedule. Post copies of your daily schedule where others can see it. That will force you to stick to it. Also schedule relaxation time for yourself! We all need that!
- Observe "good students" and make yourself look and act like a good student. Soon you will start thinking of yourself as one. We often have to "act" ourselves into a different way of thinking.

- Create a study area that meets your needs. The study area should have few distractions and be quiet. Try to keep it uncluttered. Having "a place" to study helps you get in the mood to study.
- Select the best time for you to study. We all have different "internal clocks." Study when you are most alert.
- Reflect and review daily. Reading a lesson and doing problems is not studying. You read the lesson and do the problems to gain the knowledge you need to study the material. Studying involves moving material from memory to understanding. If you can't explain a concept in your own words and relate it to something you already know, you probably don't understand it and won't remember it very long.
- Get to know others in your class and form a serious study group. Then study together at least once a week. It will keep you "honest." It is easy to slip into sloppy study habits and fool yourself into thinking you know more than you do. Having to explain ideas in a study group will bring you back to reality very quickly. We learn and learn how to learn from working with others.
- Work to stay relaxed while you are studying. For some pointers, see Study Aid 4 Learning to Relax at the end of Chapter 4.

CHAPTER 2

Just a Little Math

In this chapter and in the following chapters, we will look at the math you will need to understand chemistry. You'll be happy to learn that it is neither extensive nor extremely difficult. There are no calculus problems or quadratic equations. You do not have to plot complicated graphs. In fact the algebra you do need to know is fairly simple. Furthermore, once you set up a problem, you will do most of the computations on a calculator.

In this chapter we will review with you the mathematical concepts you need to know in order to look at answers to numerical problems and say, "Yes, this answer makes sense" or "that looks right." In the next chapter and in the rest of the book, we will show you how to do the calculations quickly and easily on a calculator..

Which Way Do I Go? Positive and Negative Numbers

In ancient times, our ancestors only used simple numbers like 1, 2 or 3. Maybe they had 3 sheep or they had 5 goats. Life was less complicated then. They did not have a symbol for "nothing" for which we now have the number "zero." Europe only got the number zero about 500 years ago from Arabic traders. Then during the Renaissance, merchants and traders also invented negative numbers to help keep track of their debts.

In mathematics we have names for these different groups of numbers. Plain old numbers like 1, 2, 3, etc., are called "natural numbers." They are called natural because they are the numbers people first used to count things. We can imagine owning 5 goats. But we can also imagine owning zero (0) goats. We can even imagine owning negative five (–5) goats, which would be the case if we owned no goats but owed 5 goats to our neighbor.

The positive numbers, the negative numbers, and zero are collectively called signed numbers. If the numbers are whole numbers, such as 5, 0, or –5, they are called integers.

You can relate positive (+) and negative (–) numbers to money you have and to money you owe. (Remember that is how negative numbers got their start.) For instance if you have $800 in the bank, you have +800 dollars or simply 800 dollars. (If a number has no sign in front of it, we assume that it is positive.) If you owe $900 on your credit card, you have a debt which you write as –900 dollars and read as "negative 900 dollars." In this case, if you add up the money you have and the money you owe, you can see that you are 100 dollars in the hole, or that you "have" –100 dollars.

The common way to portray signed numbers is with a number line:

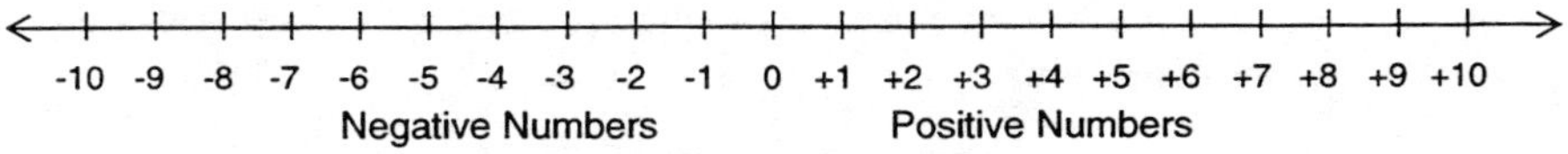

Zero is the center of the number line. The positive numbers go to the right: the negative numbers go to the left. Every positive number on the right has its opposite negative number on the left. A number that is farther to the left on the number line is smaller than a number on its right. For example, 3 is smaller than 8 so we can write

$$3 < 8$$

which we read as "3 is less than 8." If you have 3 dollars, of course, you have less money than if you have 8 dollars.

If you have an 8 dollar debt you are worse off than if you only owe 3 dollars. We can write this relationship as

$$-8 < -3$$

Notice that –8 is farther to the left on the number line than –3.

You can also relate signed numbers to the temperature scale. A thermometer is nothing but a vertical number line. Warmer temperatures, of course, are higher on the thermometer than colder temperatures. 20 degrees is warmer than, or greater than, 10 degrees.

$$20 > 10$$

But twenty degrees below zero (–20) is colder than, or less than, ten degrees below zero (–10).

$$-20 < -10$$

On a vertical number line, positive numbers go up and negative numbers go down. Positive five (+5) is five steps up from zero. Negative five (–5) is five steps down from zero.

Remember to relate new concepts to something you already know. Most of you have been balancing a checking account and reading thermometers for years, so you already know a lot about signed numbers.

Where Will You See Signed Numbers?

Three places you will routinely encounter positive and negative numbers in your chemistry class are:

- charges on subatomic particles such as protons and electrons and also on charged atoms called ions.
- exponential and scientific notation.
- temperature scales as degrees above and below zero.

Adding Signed Numbers

It is simple to add two positive numbers because it is exactly the same kind of addition you did in grade school math. Two positive numbers add to make a bigger positive number, just like we see in everyday life. Two assets add together to make a bigger asset. If Joe has 7 dollars and Mary has 8 dollars, when they combine their wealth they will have 15 dollars:

$$7 + 8 = 15$$

We can show this addition on the number line as:

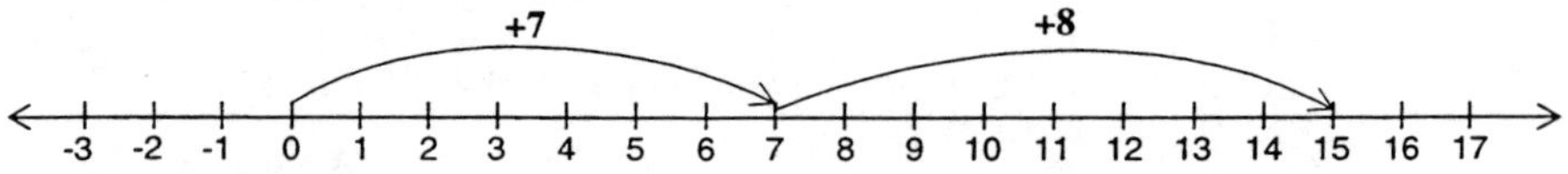

If you go 7 steps to the right of zero and then continue 8 more steps to the right, you wind up 15 steps to the right.

Two negative numbers add to make a more negative number, just as two debts add together to make an even bigger debt. Let's say Joe is \$8 in debt and Mary is \$7 in debt. When they get married they will "have" –15 dollars, that is they will be \$15 in debt:

$$-8 + (-7) = -15$$

Notice the –7 is in parentheses. We do this to separate the addition sign from the negative sign on the 7. We can show this addition on the number line as:

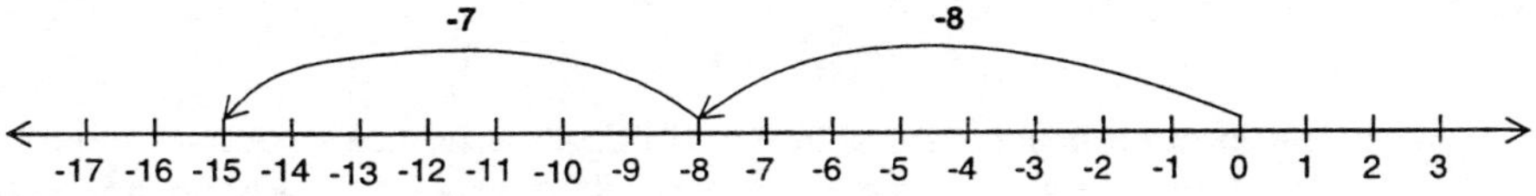

If you go 8 steps to the left of zero and then continue another 7 steps to the left, you wind up 15 steps to the left.

When you add a positive and a negative number, the sign on the answer depends on which number is larger. If Joe is seven dollars in debt and Mary has saved fifteen dollars, what do they have when they combine their resources? They end up (thanks to Mary) with 8 dollars in the bank:

$$-7 + 15 = 8$$

We can show this addition on the number line as:

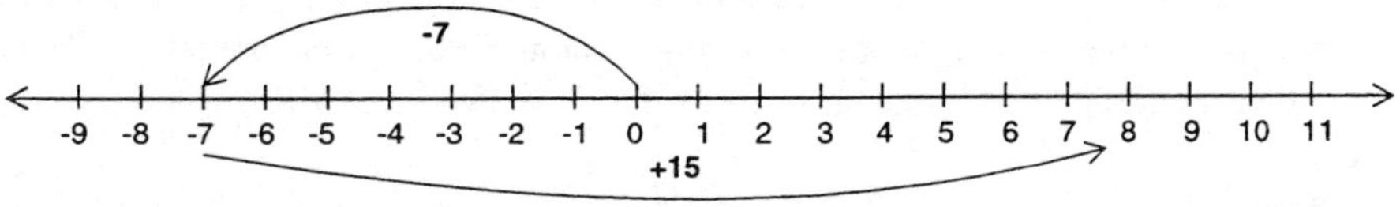

If you go 7 steps to the left of zero and then move 15 steps to the right, you wind up 8 steps to the right of zero.

When you add a positive number and a negative number, you subtract the smaller number from the larger and give the answer the sign of the larger number.

Let's look at a few more examples.

$$-6 + (-5) =$$

The signs of both numbers are negative, so add the numbers and make the sign negative.

$$-6 + (-5) = -11$$

Now try this example:

$$-10 + 13 =$$

The numbers have opposite signs, so subtract the smaller number from the larger number $(13 - 10 = 3)$. Since the sign of the larger number is positive, the answer is positive.

$$-10 + 13 = 3$$

Do one more:

$$16 + (-18) =$$

Again, the numbers have opposite signs, so we subtract the smaller number from the larger number (18 – 16 = 2). The sign of the larger number is negative, so the answer is negative.

$$16 + (-18) = -2$$

One important use of positive and negative numbers occurs when atoms form ions. Ions are atoms that have a positive or a negative charge. That charge is written in the upper right corner of the symbol for the element. For instance, you will learn that aluminum (symbol Al) forms a +3 ion which we write as Al^{+3} or Al^{3+}. Many textbooks write charges with the number before the sign (Al^{3+}), which we read as "aluminum three plus ion," rather than with the number after the sign (Al^{+3}). You will also learn that chlorine (symbol Cl) forms a -1 ion which will be written Cl^{-1} or Cl^{1-} or simply Cl^{-}. (Keep in mind that the number "one" is often omitted and merely "understood" to be there.)

Subtracting Signed Numbers

Subtracting signed numbers is a little more confusing for many people because we use the negative sign (–) in two ways: to show a number is negative and also as a subtraction or take away sign. Let's look at some examples in everyday life.

We encounter positive and negative numbers most often in our everyday life in our checking accounts. Every time you write a check, you take away a positive value from your account. Suppose you have 100 dollars deposited and you write a check for 10 dollars, you will have

$$100 - 10 = 90$$

Now suppose you don't write a check but instead the bank adds a 10 dollar service charge (–10) to your account. You now have:

$$100 + (-10) = 90$$

You can see that adding a negative 10 gives the same result as subtracting a positive 10. This same principle works in everyday life. For instance, there are two ways of disciplining children: taking away something they like (subtracting a positive) or giving them something they don't like (adding a negative).

The rule for subtraction can be summarized as: add the opposite of the second number to the first.

Let's try some examples.

$$6 - 12 =$$

We will put the positive sign on the 12 to explicitly show that the 12 is positive

$$6 - (+12) =$$

The opposite of 12 is –12, so we have

$$6 + (-12) = -6$$

Try another example.

$$-9 - 18 =$$

First rewrite the problem showing that the 18 is positive.

$$-9 - (+18) =$$

The opposite of 18 is –18, so we add

$$-9 + (-18) = -27$$

If you already owed nine dollars and someone took your credit card and gave you an eighteen dollar debt, it would have the same result as if that person took eighteen dollars from you.

Now suppose you owe sixteen dollars on your credit card. Then the bank tells you that it made a mistake and is taking away $9 of the debt. We can write this as

$$-16 - (-9) =$$

The opposite of –9 is 9, so we add

$$-16 + 9 = -7$$

You would then only owe $7.

Taking away a negative is comparable to adding positive. Another analogy from everyday life might help you understand: if a doctor removes a pain or a disease (a negative), the patient feels better (more positive).

There is another way to look at the subtraction of signed numbers. We can look at subtracting as finding how far apart two numbers are, which is called finding the difference of the numbers. Let's look at an example. If the temperature changed from 70° in the afternoon to 45° in the evening, how much did the temperature change from afternoon to evening? Finding the difference of two numbers always follows this pattern:

$$\text{Difference} = \text{To} - \text{From}$$

We can write this problem as a subtraction

$$45 - 70 =$$

$$45 + (-70) = -35$$

The temperature became 35 degrees colder in the evening.

I (Tim) once took a train trip across Canada in the middle of January from Winnipeg, where it was $-30\,^{\circ}\mathrm{F}$, to Vancouver, where it was $50\,^{\circ}\mathrm{F}$. What was the difference in temperature?

$$\text{Difference} = \text{To} - \text{From}$$

$$50 - (-30) = 50 + 30 = +80$$

It was 80 degrees warmer in Vancouver than it was in Winnipeg.

In the chapter on radioactivity you will study how radioactive atoms lose beta (β) particles. The nucleus of an atom has positive particles called protons and neutral particles with a zero charge called neutrons. When a radioactive nucleus becomes unstable, a neutron loses a negative particle called a beta particle. For instance when a carbon nucleus with a charge of 6 loses a beta particle with a -1 charge, its charge changes to 7.

$$+6 - (-1) = +7$$

$$6 + 1 = 7$$

Take away a negative 1 from 6 and it becomes 7. So when a nucleus loses a negatively charged beta particle, its charge always increases by one.

Multiplying and Dividing Signed Numbers

When you multiply two numbers or "take their product," the problem can be written several different ways:

$$2 \times 3 = 6 \qquad 2 \bullet 3 = 6 \qquad 2 * 3 = 6 \qquad 2(3) = 6$$

The form you will see most often is:

$$2(3) = 6 \qquad \text{or} \qquad (2)(3) = 6$$

where the numbers are simply written side-by-side separated by parentheses.

The product of two positive numbers is a positive number which you already know from grade school math.

$$2(10) = 20 \qquad 3(10) = 30 \qquad 4(10) = 40$$

The product of a positive number and a negative number is a negative number. If you double or triple, etc. a debt, you will have an even bigger debt.

$$2(-10) = -20 \qquad 3(-10) = -30 \qquad 4(-10) = -40$$

The rule that many people find a little strange at first is that the product of two negative numbers is a positive number. But if you think about it in ordinary language, the rule makes sense. If I take away two (–2) of your $10 debts (–10), you will be $20 richer (+20), if I take away 3 of your $10 debts, you will be $30 richer, etc.

$$-2(-10) = 20 \qquad -3(-10) = 30 \qquad -4(-10) = 40$$

It's important you do not mix up the rules for multiplication of signed numbers with the rules for addition.

$-3 + (-5) = -8$ negative + negative = negative

but

$-3(-5) = 15$ negative × negative = positive

There is one multiplication you will see that may not look like a multiplication. If you see a negative sign in front of a set of parentheses, it means you multiply by negative one.

$$-(-5) \text{ means } -1(-5) = 5$$

$$-(-11) \text{ means } -1(-11) = 11$$

Note that multiplying by –1 simply changes the sign of the number.

The rules for division are exactly the same as for multiplication.

A positive number divided by a positive number gives a positive answer:

$$12 \div 2 = 6$$

A negative number divided by a positive number or a positive number divided by a negative number gives a negative answer:

$$(-12) \div 2 = -6 \qquad \text{or} \qquad 12 \div (-2) = -6$$

A negative number divided by a negative number yields a positive answer:

$$(-12) \div (-2) = 6$$

Picking up the Pieces
Fractions, Decimals and Percents

Fractions

We use fractions to show something is cut into equal pieces. The bottom of the fraction, called the denominator, tells us how many parts the whole is cut into. The top of the fraction, called the numerator, tells us how many of the parts we have. The line that separates them is called the fraction line or fraction bar.

$$\frac{\text{numerator}}{\text{denominator}}$$

The fraction $\frac{4}{5}$ tells us that something is cut into five equal pieces, and we have four of them. It might represent a pie cut into five equal pieces and there are four pieces left (Figure 2.1) or it could be your grade if you got four questions right on a five question quiz.

A fraction can be expressed in words in several ways. We can say four out of five (as in 4 out of 5 doctors recommend...), four-fifths, four per five, 4 divided by 5 or as the ratio of 4 to 5.

Notice that $\frac{4}{5}$ of a pie is less than one whole pie. Any fraction smaller than one is called a proper fraction. Before a piece of the pie was eaten, we had $\frac{5}{5}$ or one whole pie. (Remember a fraction is another way of writing a division; therefore $\frac{5}{5}$ means $5 \div 5 = 1$.)

Any number or variable or unit divided by itself equals one.

Examples: $\frac{5}{5} = 1$ $\frac{10}{10} = 1$ $\frac{x}{x} = 1$ $\frac{\text{inches}}{\text{inches}} = 1$

Fractions that are equal to one or larger, such as $\frac{9}{5}$, are called improper fractions.

You will see the word "per" often in chemistry problems. Any time you see "per," it means you have a fraction. If you drive 450 miles *per* 12 gallons of gasoline, you can write it as the fraction:

$$\frac{450 \text{ miles}}{12 \text{ gallons}}$$

The item following per is always in the denominator, the bottom of the fraction. This example also shows you the type of fraction you will see most often in chemistry, which will not be just plain numbers but will have "units" like gallons or miles.

Likewise if we drive 65 miles per hour, we can write the fraction:

$$\frac{65\text{ miles}}{1\text{ hour}}$$

Notice that if we have a fraction with units such as 65 miles per hour, we can write it as 65 miles/hour, $\frac{65\text{ miles}}{\text{hour}}$ or $\frac{65\text{ miles}}{1\text{ hour}}$ where the third form emphasizes that a simple number such as 65 can also be written as a fraction with a denominator of 1. Then we have both a number and a unit in both numerator and denominator. We will talk more about fractions with units in Chapter 5.

Fractions are often written in textbooks with a slanted fraction line (/), mainly because it only takes one line of type. Thus 4/5 is the same as $\frac{4}{5}$. To make a fraction with a slanted line easier to work with, you may want to rewrite it with a horizontal fraction line.

Multiplication is the main operation you will perform with fractions in chemistry. Here are some examples of multiplying fractions:

$$\frac{\overset{1}{\cancel{2}}}{\underset{1}{\cancel{3}}} \times \frac{\overset{1}{\cancel{3}}}{\underset{8}{\cancel{16}}} = \frac{1}{8} \qquad \frac{\overset{1}{\cancel{4}}}{5} \times \frac{3}{\underset{1}{\cancel{4}}} = \frac{3}{5} \qquad \overset{20}{\cancel{2000}} \times \frac{1}{\underset{1}{\cancel{100}}} = 20$$

Notice how the numbers can be reduced before multiplying.

When you see a statement like “1/4 of” or “1/2 of”, it means you are going to multiply.

$$\frac{1}{4} \text{ of } \frac{1}{2} \text{ means } \frac{1}{4} \times \frac{1}{2} = \frac{1}{8}$$

If we cut 1/2 of a pie into 4 pieces, each piece will be 1/8 of the whole pie.

Likewise

$$\frac{1}{3} \text{ of } 9 \text{ means } \frac{1}{\cancel{3}} \times \frac{\overset{3}{\cancel{9}}}{1} = \frac{3}{1} = 3$$

One-third of nine dollars is three dollars. Notice that multiplying by the fraction 1/3 gives the same result as dividing by 3.

We can take any fraction or whole number and write its reciprocal. The reciprocal of a fraction is simply the fraction turned upside down.

The reciprocal of $\frac{9}{5}$ is $\frac{5}{9}$. The reciprocal of $\frac{7}{9}$ is $\frac{9}{7}$. The reciprocal of $\frac{1}{100}$ is $\frac{100}{1}$ or 100.

If we want to take the reciprocal of a whole number such as 100, we can always write the number with a denominator of 1, for instance $100 = \frac{100}{1}$. Then to find the reciprocal of 100, we just turn its fraction form upside down to get $\frac{1}{100}$.

If the fraction has units, we must move the units along with their numbers. The reciprocal of $\frac{5280 \text{ feet}}{1 \text{ mile}}$ is $\frac{1 \text{ mile}}{5280 \text{ feet}}$.

How many quarters are there in a \$10 roll of quarters? We can write this as a division problem:

$$10 \div \frac{1}{4} = 40 \qquad \text{or} \qquad \frac{10}{\frac{1}{4}} = 40$$

which we can state in words as "How many $\frac{1}{4}$'s are there in 10?" You know from experience that there are 4 quarters in a dollar. So we can also write this problem as a multiplication

$$10 \times 4 = 40$$

You can see that multiplying by 4 (the reciprocal of $\frac{1}{4}$) gives the same result as dividing by $\frac{1}{4}$.

In general, to divide a number by a fraction, multiply the number by the reciprocal of the fraction. Let's look at a couple examples:

$$10 \div \frac{1}{10} = 10 \times \frac{10}{1} = 10 \times 10 = 100$$

$$\frac{3}{5} \div \frac{3}{10} = \frac{\overset{1}{\cancel{3}}}{\underset{1}{\cancel{5}}} \times \frac{\overset{2}{\cancel{10}}}{\underset{1}{\cancel{3}}} = 2$$

How many pennies are there in one dollar? Of course there are one hundred. We can write this as a division problem:

$$1 \div \frac{1}{100} = 1 \times \frac{100}{1} = 1 \times 100 = 100$$

How many quarter slices are there in one pie?

$$1 \div \frac{1}{4} = 1 \times \frac{4}{1} = 1 \times 4 = 4$$

Can you see the pattern? When you divide something into $\frac{1}{100}$'s, you have 100 pieces. When you cut something into $\frac{1}{4}$'s, you have 4 pieces. We will see this pattern again when we look at the metric system in Chapter 4.

In the metric system you will see this same pattern. One tenth of a meter is called a decimeter. If each decimeter is 1/10 of a meter, the meter has been cut into 10 equal pieces. So we can say:

1/10 meter = 1 decimeter   or   1 meter = 10 decimeters

Decimals

Often decimals are more convenient to use than fractions, especially when we use a calculator. Here is a decimal place value chart to refresh your memory.

thousands	hundreds	tens	ones	decimal point	ten*ths*	hundred*ths*	thousand*ths*	ten-thousand*ths*	hundred-thousand*ths*	million*ths*	ten-million*ths*	hundred-million*ths*	billion*ths*
3	7	3	5	.	2	5	7	8	2	6	4	2	9

You read the decimal 0.539 as five hundred thirty-nine thousandths which is equivalent to the fraction $\frac{539}{1000}$. The decimal 0.69 means sixty-nine hundredths and is equivalent to the fraction $\frac{69}{100}$. The decimal 0.000007 means seven millionths and is the same as the fraction $\frac{7}{1000000}$. You don't have to write the zero in front of the decimal point, but most of the time we put it there to help keep the decimal point from getting lost.

To change a fraction into a decimal, simply divide the denominator into the numerator.

$$\frac{3}{4} = 0.75 \qquad \frac{9}{5} = 1.8 \qquad \frac{4}{5} = 0.8 \qquad \frac{1}{3} = 0.333$$

In the first three examples above the division stops or terminates. In the fourth example the 3's repeat endlessly and we have to round the answer off. We will discuss rounding off in detail in the next chapter.

In the metric system you will often see fractions with multiples of ten in the denominator. A "short cut" for dividing these fractions is to move the decimal point of the numerator to the left as many places as the number of zeros in the denominator.

$$\frac{1}{10} = 0.1 \qquad \frac{1}{100} = 0.01 \qquad \frac{1}{1000} = 0.001 \qquad \frac{1}{1{,}000{,}000} = 0.000001$$

In the next chapter we will discuss multiplying and dividing decimals using a calculator. But you can multiply and divide decimals by multiples of ten in your head. To multiply a decimal by a multiple of 10, move the decimal point to the right a number of places equal to the number of zeros.

$0.485 \times 1000 = 485$ move 3 places

$0.00057 \times 100 = 0.057$ move 2 places

$25 \times 10 = 250$ move 1 place

To divide a decimal by a multiple of 10, move the decimal point to the left a number of places equal to the number of zeros.

$\frac{0.485}{1000} = 0.000485$ move 3 places

$\frac{0.000057}{100} = 0.00000057$ move 2 places

$\frac{25}{10} = 2.5$ move 1 place

Percents

Percents are special fractions that mean "out of one hundred." "Per" means "out of" and "cent" means one hundred, just as there are one hundred cents in a dollar and one hundred years in a century. Here are some examples demonstrating the relationships among percents, fractions, and decimals:

$$9\% = \frac{9}{100} = 0.09 \qquad 89\% = \frac{89}{100} = 0.89$$

Notice that, because we are dividing by 100 when we convert a percent to a decimal, we move the decimal point two places to the left.

If you have percents less than 1, they are handled the same way. Normal saline salt solution that is used in hypodermic injections is 0.9% salt. We can convert this percent to a decimal like this:

$$0.9\% = \frac{0.9}{100} = 0.009$$

In computing taxes, a *mill* is 0.1% of a dollar. We can convert this percent to a decimal:

$$0.1\% = \frac{0.1}{100} = 0.001$$

So a 1-mill levy will tax you 1 dollar out of every thousand dollars.

What is 100%? One hundred percent is one, the whole thing.

$$100\% = \frac{100}{100} = 1$$

So if you give something your whole effort, you are giving 100%.

We use percents when we want to have an easy way to see how big the part is compared to the whole. Thus if you answered 45 problems right out of a possible 60 on one test and 43 right out of 50 on the second test, it is not immediately obvious that you actually did better on the second test. On the first test you received

$$\frac{45}{60} \times 100\% = 75\%$$

while on the second test you received

$$\frac{43}{50} \times 100\% = 86\%$$

In general percents are calculated

$$\text{Percent} = \frac{\text{part}}{\text{whole}} \times 100\%$$

In glucose (a sugar), 72 grams out of 180 grams are carbon. What is the percent of carbon?

$$\text{Percent} = \frac{\text{part}}{\text{whole}} \times 100\% = \frac{72\text{ gram}}{180\text{ gram}} \times 100\% = 40\%$$

Suppose 180 students out of 200 passed the chemistry course. What percent passed?

$$\text{Percent} = \frac{\text{part}}{\text{whole}} \times 100\% = \frac{180\text{ students}}{200\text{ students}} \times 100 = 90\%$$

There is one other type of percent problem you may see. In this problem you are given a percent and a total and you want to find the part. Let's look at some examples.

If 20% of 2500 students are from out-of-state, how many students are from out-of-state? Remember 20% means the fraction $\frac{20}{100}$ and when we say "of," it means to multiply. So we write:

$$20\% \times 2500 = \frac{20}{100} \times 2500 = 500 \quad \text{or} \quad 0.20 \times 2500 = 500$$

In general we can write

$$\text{Part} = \frac{\%}{100} \times \text{Total} \qquad \text{or} \qquad \text{Part} = (\text{Percent as a decimal}) \times \text{Total}$$

What is 35% of 120?

$$\frac{35}{100} \times 120 = 42 \quad \text{or} \quad 0.35 \times 120 = 42$$

If you want to leave a 15% tip on a $55 meal, how much do you leave?

$$\text{Part} = \frac{15}{100} \times 55 = 8.25 \quad \text{or} \quad 0.15 \times 55 = 8.25$$

So you leave a little over $8.00.

The Little Big Numbers
Exponents

Whenever we multiply the same number by itself many times, we can shorten the writing required by using exponents. If we multiply 5 four times

$$5 \times 5 \times 5 \times 5$$

we can write it as

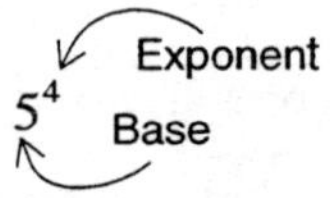

The 5 is the base; the 4 is the exponent. In words, we say 5 to the fourth power.

In chemistry numbers are often written in powers of 10 with a base of "10" like this:

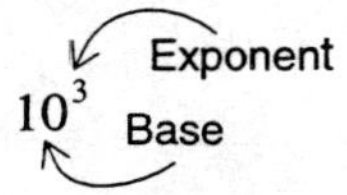

We read 10^3 as 10 to the third power or 10 cubed,

$$10^3 = 10 \times 10 \times 10 = 1000$$

Now look at the following pattern:

$$10^1 = 10$$
$$10^2 = 10 \times 10 = 100$$
$$10^3 = 10 \times 10 \times 10 = 1000$$
$$10^4 = 10 \times 10 \times 10 \times 10 = 10{,}000$$
$$10^5 = 10 \times 10 \times 10 \times 10 \times 10 = 100{,}000$$
$$10^6 = 10 \times 10 \times 10 \times 10 \times 10 \times 10 = 1{,}000{,}000$$

You can see from the pattern that the exponent is equal to the number of zeros. Once you get used to exponents, it is much easier to write 10^{12} than it is to write 1,000,000,000,000. It will be helpful to you to become familiar with the names of the exponential forms of key powers of ten because they are used so often.

$$10^3 = 1 \text{ thousand} = 1000$$
$$10^6 = 1 \text{ million} = 1{,}000{,}000$$
$$10^9 = 1 \text{ billion} = 1{,}000{,}000{,}000$$
$$10^{12} = 1 \text{ trillion} = 1{,}000{,}000{,}000{,}000$$

Negative Exponents

Exponents can also be negative. A negative exponent means to take the reciprocal of the base and give the base a positive exponent. The negative exponents of 10 follow this pattern:

$$10^{-1} = \frac{1}{10} = 0.1$$

$$10^{-2} = \frac{1}{10^2} = \frac{1}{100} = 0.01$$

$$10^{-3} = \frac{1}{10^3} = \frac{1}{1000} = 0.001$$

$$10^{-4} = \frac{1}{10^4} = \frac{1}{100{,}000} = 0.0001$$

$$10^{-5} = \frac{1}{10^5} = \frac{1}{100{,}000} = 0.00001$$

$$10^{-6} = \frac{1}{10^6} = \frac{1}{1{,}000{,}000} = 0.000001$$

$$10^{-9} = \frac{1}{10^9} = \frac{1}{1{,}000{,}000{,}000} = 0.000000001$$

For negative exponents, the exponent is equal to the number of decimal places. Negative exponents are convenient for small numbers. Compare 10^{-12} to 0.000000000001! Keep in mind the larger the negative exponent, the smaller the number actually is.

One very important exponent is zero. $10^0 = 1$. In mathematics anything to the zero power is 1. Thus $2^0 = 1$; $x^0 = 1$; etc.

To get an idea why, let's continue the pattern for zeros on the last page:

$10^4 = 10{,}000$	4 zeros
$10^3 = 1000$	3 zeros
$10^2 = 100$	2 zeros
$10^1 = 10$	1 zero
$10^0 = 1$	no zeros

We will show you another reason why $10^0 = 1$ shortly.

Multiplying and Dividing Exponential Numbers

When you multiply two exponential numbers with the same base, you simply add the exponents. Let's look at how this works:

$$10^3 \times 10^4 = (10 \times 10 \times 10) \times (10 \times 10 \times 10 \times 10) = 10^{3+4} = 10^7$$

As you can see, a total of seven 10's are being multiplied. Adding the exponents is the shortcut.

Note: the multiplication sign between the parentheses is often omitted and the problem written

$$(10 \times 10 \times 10)(10 \times 10 \times 10 \times 10) =$$

This rule works whether the exponents are positive or negative. We'll do some examples using the rules we learned in the last section for adding signed numbers.

$$10^{-3} \times 10^{-9} = 10^{-3+(-9)} = 10^{-12}$$

$$10^{-4} \times 10^{8} = 10^{-4+8} = 10^{4}$$

$$10^{-8} \times 10^{-8} = 10^{-8+(-8)} = 10^{-16}$$

$$10^{-9} \times 10^{23} = 10^{-9+23} = 10^{14}$$

Now let's see if we can prove to ourselves that 10^0 really is equal to 1.

$$10^{-1} \times 10^{1} = 10^{-1+1} = 10^{0}$$

or, if we convert these exponential numbers to fractions we get

$$\frac{1}{\underset{1}{\cancel{10}}} \times \frac{\overset{1}{\cancel{10}}}{1} = 1$$

So 10^0 must equal 1.

When we divide exponential numbers, we subtract the exponent of the denominator from the exponent of the numerator. Let's see how this works. (Remember that any number divided by itself is equal to one. 10/10 = 1)

$$\frac{10^4}{10^7} = \frac{\cancel{10} \times \cancel{10} \times \cancel{10} \times \cancel{10}}{\cancel{10} \times \cancel{10} \times \cancel{10} \times \cancel{10} \times 10 \times 10 \times 10} = \frac{1}{10 \times 10 \times 10} = \frac{1}{10^3} = 10^{-3}$$

The four 10's in the numerator cancel four of the seven 10's in the denominator. We get the same result by subtracting the exponents

$$10^{4-7} = 10^{4+(-7)} = 10^{-3}$$

Let's try some more:

$$\frac{10^{-3}}{10^{-4}} = 10^{-3-(-4)} = 10^{-3+4} = 10^{1} = 10$$

$$\frac{10^{-3}}{10^{4}} = 10^{-3-4} = 10^{-3+(-4)} = 10^{-7}$$

$$\frac{10^{6}}{10^{9}} = 10^{6-9} = 10^{6+(-9)} = 10^{-3}$$

Chapter 2

If You Get Tired of Counting Zeros
Scientific Notation

You will use scientific notation often in your chemistry class. A number written in scientific notation is simply an ordinary decimal number between 1 and 10, called the coefficient, multiplied by a power of ten.
Here is an example:

$$5.24\times10^{4}$$

Power of 10

Coefficient

The decimal form of 5.24×10^{4} is 52,400 because 5.24×10^{4} is $5.24 \times 10{,}000 = 52{,}400$

The exponent can also be negative:

$$2.39\times10^{-4}$$

The decimal form of 2.39×10^{-4} is 0.000239 because we can write 10^{-4} as 0.0001 and multiply:

$$2.39\times10^{-4} = 2.39 \times 0.0001. = 0.000239$$

In the next chapter you will learn to multiply and divide using scientific notation as well as convert between scientific notation and decimal form quickly and easily using your calculator.

But now let's convert some numbers from decimal form to scientific notation without a calculator just so we can get a feel for the numbers. To convert a decimal number to scientific notation, move the decimal point until you get a number between 1 and 10. The exponent will be the number of places you moved the decimal point. The sign of the exponent will depend on whether it is a small number (negative exponent) or a big number (positive exponent).

$$120{,}000 = 1.2\times10^{5}$$

5 places

+ exponent: large number

If you don't see a decimal point, assume it comes at the end of the number.

$$0.000027 = 2.7\times10^{-5}$$

5 places

– exponent: small number

$1{,}870{,}000{,}000 = 1.87\times10^{9}$ (9 places) — + exponent: large number

$0.0000000905 = 9.05\times10^{-8}$ (8 places) — – exponent: small number

Scientific notation puts unwieldy large or small numbers into a form that is much easier to use. Sometimes you will see regular size numbers written in scientific notation.

$$18 = 1.8\times10^{1}$$

$$0.19 = 1.9\times10^{-1}$$

As you can see, numbers like these are easier to write in decimal form.

To convert a number written in scientific notation to decimal form, move the decimal point of the coefficient a number of places equal to the exponent, adding zeros as necessary. If the exponent is positive, move the decimal point to the right to make a large number. If the exponent is negative, move the point to the left to make a small number. Let's look at some examples:

$2.37\times10^{5} = 237{,}000$ — move 5 places to the right

$6.59\times10^{-4} = 0.000659$ — move 4 places to the left

Some Practice Problems *Answers on page 72*

1. $8 + (-4) =$
2. $8 - (-4) =$
3. $8(-4) =$
4. $\frac{8}{-4} =$
5. $-8 + (-4) =$
6. $-8 - (-4) =$
7. $-8(-4) =$
8. $\frac{-4}{-8} =$
9. $\frac{1}{2}$ of $\frac{1}{3} =$
10. $\frac{1}{2} \div \frac{1}{3} =$
11. $\frac{1}{2}$ of $8 =$
12. $\frac{1}{2} \times 8 =$

13. $\frac{1}{2} \div 8 =$

14. $8 \div \frac{1}{2} =$

15 Convert $\frac{1}{2}$ to a decimal.

16. Convert $\frac{11}{5}$ to a decimal.

17. Convert $\frac{3}{10}$ to a decimal.

18. Convert $\frac{15}{100}$ to a decimal.

19. Convert 13% to a decimal.

20. Convert 1.3% to a decimal.

21. If you get 40 out of 50 problems correct on a test, what is your percent grade?

22. If you want to score 90% on a 60 problem test, how many problems do you need to get correct?

23. $10^5 \times 10^8 =$

24. $\frac{10^5}{10^8} =$

25. $10^{-2} \times 10^{-1} =$

26. $\frac{10^{-2}}{10^{-1}} =$

27. Convert 8×10^7 to decimal form.

28. Convert 1.3×10^{-3} to decimal form.

29. Convert 289,000 to scientific notation.

30. Convert 0.000054 to scientific notation.

Study Aid 2 Reading Your Textbook

If your assignment is to read pages 88-108, the worst thing you can possibly do, aside from not doing the reading at all, is to open the text to page 88 and begin reading. Reading an assignment is a three-phase process. It has a BEFORE, DURING, AND AFTER phase and each is equally important!

During the BEFORE phase:

- Skim the entire assignment first so you have a general idea of the content and try to relate the content to things you already know.
- Look at your syllabus and see how many lectures are devoted to the assignment. See what other material is included with this assignment on the next test.

- Judging from the length and the difficulty of the reading assignment, try to determine about how long the task will take and plan accordingly. (Remember this is technical reading, so reading only 1-2 pages per hour is not unreasonable.)

During the DURING phase:

- Mark up your text. Write notes in the margin. Highlight key concepts. (Try not to buy a book that has been marked up too much. You will be distracted by someone else's markings.)
- If examples do not show all the steps, write in the missing steps so you will remember how to do the problem when it is time to review for a quiz or test.
- Alice in Wonderland asked, "What good is a book without pictures?" Fortunately most modern chemistry books are filled with pictures. Take advantage of them. Remember, chemistry is visual. When the text refers to a drawing, figure, diagram, chart, photograph, etc., look at it right away. Study it. Sometimes one picture can be worth many words!
- Spend time reflecting on what you read. Can you put the concepts into your own words?
- Do you need to outline, make a diagram or write a summary of the material?
- Do you need to hear yourself reading the material? Reading difficult material aloud slows you down and lets you hear as well as see. Two senses are better than one! Remember, you are reading for understanding.

During the AFTER phase:

- Go back to any sections you did not understand. If after going over them again, you still have concerns, put a question mark beside them and ask about them during the next class or help session or meeting with your study group.
- Go back over the entire reading assignment looking at main headings and the notes you made in the margin. Does this information remind you of the main ideas and content of each section? If not, add to it.
- Write a brief summary of what you read. Include any type of diagram that will help you remember the material. This summary, what you have underlined in the text, what you have written in the margin in the text, and the notes you have taken in class are the materials you need to use when you review daily and study for quizzes and tests.

CHAPTER 3

Push Button Thinking Using Your Calculator

Probably the best friend you have in a chemistry class is your calculator. If you use your calculator correctly, it will save you precious seconds during a test. It will make doing homework problems outside class much more efficient and enjoyable, and allow you to spend more time on the concepts rather than laboring over the calculations. It is much more important to understand the concepts of chemistry than to do the calculations.

Ten Dollar Calculators That Work

Many students have difficulty in the first weeks of the semester because they don't have the type of calculator they need to quickly and efficiently perform the tasks the instructor and textbook will require of them. You may sit down in your first class with a four-function calculator that just does the basic operations of adding, subtracting, multiplying and dividing—the type they give you when you open a checking account. This type of calculator is not powerful enough to do many of your chemistry problems. On the other hand, you don't want to purchase an engineering calculator with so many functions it is intimidating to look at—and has a manual as thick as *War and Peace*!

You need a scientific calculator that will help you quickly and easily dispatch the problems set before you in your class, especially on tests. The calculator should have the following capabilities:

- The four basic arithmetic functions.
- A button to enter scientific notation ([EE] or [EXP] on most calculators).

- Buttons to easily enter logarithms. (Don't worry about what logarithms are right now. We will discuss these later in Chapter 16.)
- Buttons to convert between scientific notation and decimal form.
- Buttons to do statistical calculations. (You won't need them in this class, but many of you will probably take statistics later in your program. You may also need to do some simple statistics in some of your more advanced nursing classes.)
- At least 10 places on the display so you can display fairly large numbers in decimal form. (Called 10 + 2 on some models.)

Below is a list of calculators that are inexpensive (under $10) and do the job you need a calculator to do.

- Casio fx-250HA or fx-250HB
- Texas Instruments TI-30X or TI-30X Solar

The First Five Buttons
The Basic Operations

Nothing is easier than multiplying or dividing on a calculator. If you want to multiply 2.71 × 1.91, you simply punch

2.71 [×] 1.91 [=] The display reads **5.1761**

When you divide two numbers, the process is the same. Let's divide 3.75 by 1.2. This problem can be presented to you as 3.75 ÷ 1.2 or 3.75/1.2 or $\frac{3.75}{1.2}$. No matter how the division is written, it is entered into the calculator as

3.75 [÷] 1.2 [=] The display reads **3.125**

Most of the time in chemistry you will do several operations in one problem. An example of this type of problem is

$$2.875 \times \frac{1}{3.125} =$$

Many students enter the operations like this:

2.875 [×] 1 [÷] 3.125 [=] The display reads **0.92**

It is much simpler and less prone to error if the multiplication step is left out when all you are doing is multiplying by one. You simply enter

2.875 [÷] 3.125 [=] The display reads **0.92**

Later, when you do conversions, you will do many multiplications and divisions in one problem. Let's try one:

$$\frac{2.375 \times 4.15}{0.03125 \times 1.577} =$$

Many students do the problem like this:

First they multiply the numbers in the numerator

2.375 [×] 4.15 [=] The display reads **9.85625**

Then they write this answer down, push [C] to clear the display and then multiply the numbers in the denominator:

0.03125 [×] 1.577 [=] The display reads **0.04928125**

Then they write this number down, clear the display, reenter 9.85625, and finally divide by 0.04928125 to get the answer 200.

In other words, they multiply the numbers in the numerator first, erase the answer, then multiply the numbers in the denominator, reenter the first answer and then do the division.

There is an easier way to do it. Do the multiplications and divisions in order from left to right. Press [×] before numbers that are in a numerator (after the first number) and press [÷] before any number in a denominator.

2.375 [×] 4.15 [÷] 0.03125 [÷] 1.577 [=] The display reads **200**

You can change the order, as long as you press [×] before a number in the numerator and press [÷] before a number in the denominator.

2.375 [÷] 0.03125 [÷] 1.577 [×] 4.15 [=] The display reads **200**

Let's try another:

$$14.4 \times \frac{1}{1.6} \times \frac{13.3}{20} =$$

We punch:

14.4 [÷] 1.6 [×] 13.3 [÷] 20 [=]   The display reads   **5.985**

Doing problems this way can save you much time on test questions.

What Do I Do With All These Numbers? Rounding Off and Significant Figures

If you divide 2 by 7 on your calculator, the display reads 0.285714286. Now ten decimal places can be quite cumbersome. Fortunately, as you know, you do not have to write all those numbers when you answer a problem. You can round off your answer. Answers beyond a certain number of places are not just cumbersome, they are misleading. So we round our answers off to a reasonable number of decimal places.

In order to round off, we have to understand significant figures (sometimes referred to as sig figs). In chemistry (and in all science) we do not work with pure numbers like in algebra class. We work with measurements. If you take your body temperature with a fever thermometer, it may look like the figure below. The mercury stops between 98 and 99 degrees. It looks like it is between 98.2 and 98.4 degrees (each small line on the thermometer is 0.2 of one degree), so we might guess 98.3. In all measurements the last digit is a good guess and your guess might be a little different from mine. This temperature (98.3 degrees) has three significant figures. If we round it off to two places, it will have two significant figures (98 degrees).

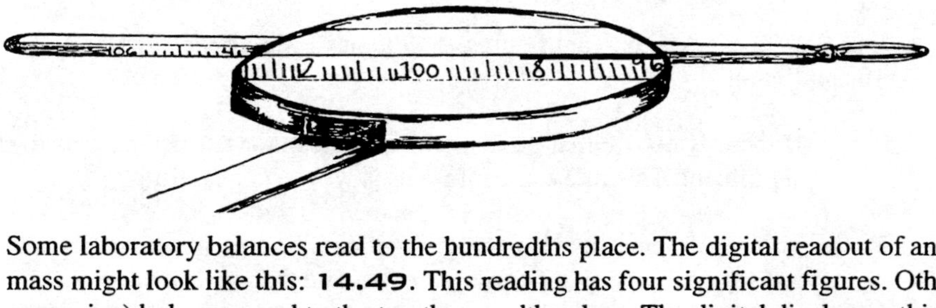

Some laboratory balances read to the hundredths place. The digital readout of an object's mass might look like this: **14.49**. This reading has four significant figures. Other (more expensive) balances read to the ten-thousandths place. The digital display on this balance might read **14.4927** for this object. This reading has six significant figures.

What Are the Rules for Significant Figures?

How can you tell if digits are significant? Most of the confusion involves the number zero. Your textbook will give you 4 or 5 rules something like these:

1. All non-zero digits are significant.

This means that any number that is not zero is significant. 7.121486 has 7 significant figures; 4215 has 4 significant figures.

2. All zeros between non-zero digits are significant.

7002 has 4 significant figures; 7.1002 has 5 significant figures; 702 has 3 significant figures; 80.003 has 5 significant figures.

3. All zeros to the left of the first non-zero digit are not significant.

Here is where your normal everyday definition of "significant" may clash with the chemistry definition. The number 0.0025 has only 2 significant figures, as the zeros to the left of the 2 are not significant. That does not mean they don't have to be there. If you took them away, you would get .25, a very different number! A common dosage of Vasotec, a blood pressure-lowering drug, is 0.0025 grams. If you ignored the zeros, the dosage would become .25 grams, 100 times too high and probably fatal to the patient. Those zeros then have the very important job of being place holders; however, they are not significant in the sense of expressing the accuracy of the number.

If we convert 0.0025 kilograms to grams, it becomes 2.5 grams. Because we changed the units, we no longer needed the left-hand zeros as place holders.

4. When a decimal point is present, zeros to the right of the last non-zero digit are significant.

Thus 0.2300 has 4 significant figures, 2.4900 has 5 significant figures and 2.0 has 2 significant figures.

5. If there is no decimal point written, you cannot tell if the zeros to the right are significant. This rule can confuse us if we are not careful.

For instance, we cannot say how many significant figures whole numbers like 8000 and 400 have. They have one significant figure; beyond that we can not say.

Now let's look at some examples that involve all the rules.

7.0010 has 5 significant figures. (All the zeros are significant because two are between non-zero digits and the last is in a number that has the decimal point written.)

0.0005 has 1 significant figure. (The zeros to the left are not significant.)

0.000500 has 3 significant figures. (The zeros to the left of the 5 are place holders but are not significant. The zeros to the right are significant because a decimal point is shown.)

1.000500 has 7 significant figures. (All the zeros are significant because the three zeros are between non-zero digits and the last two are to the right in a number that has the decimal point written.)

7000 has 1 significant figure. (No decimal point is shown so you cannot tell whether or not the zeros are significant.)

7000. has 4 significant figures. (This time the decimal point is shown.)

Scientific Notation to the Rescue

Let's look at the number 800. Suppose it represents 800 pounds. If 800 pounds is just a good guess at the weight of an object, then the number has only 1 significant figure. However, if you weighed the object and it weighed exactly 800 pounds, the number would have 3 significant figures. If you want to show that the number has 3 significant figures, you would have to write it with a decimal point included, that is, 800. lbs. But it is easy for a decimal point at the end of a whole number to "get lost." However, we can easily show the number of significant figures in numbers such as this by converting them to scientific notation. If you convert 800 to scientific notation, it becomes

$$8 \times 10^2$$

If you want to show that it has 3 significant figures, you can write it as

$$8.00 \times 10^2$$

When you write a number in scientific notation, the number of significant digits is expressed by the coefficient.

7000 lbs becomes	7×10^3 with 1 significant figure
7000 lbs becomes	7.0×10^3 with 2 significant figures
7000 lbs becomes	7.00×10^3 with 3 significant figures

7000 lbs becomes 7.000×10^3 with 4 significant figures

Remember that the exponent plays no part in determining the number of significant figures. It merely tell us where to put the decimal point.

Rules for Rounding Off

You may remember these rules from a math class in high school. To round off a number to a certain place, you look at the number in the next place to its right. If the number to its right is less than 5, that is, if it is 0, 1, 2, 3, or 4, you leave the number alone. This is called rounding down. If the number to its right is 5 or greater, that is, if it is 5, 6, 7, 8, or 9, you increase the value of the number by one. This is called rounding up.

Let's round some numbers.

7.134 rounded to the hundredths place becomes 7.13 since the number in the thousandths place is 4, a number less than 5.

0.21029 rounded to the ten-thousandths place becomes 0.2103 since the number to the right of that place is 9, a number that is 5 or greater.

Sometimes when you round a number off it makes it a little easier to visualize the size of the number. The concentration of salt in normal saline solution, used in hypodermic solutions, is 0.9%. If we round the percent off to the one's place, it becomes 1% or about 1 part of salt in one hundred parts of solution.

Periodic tables in chemistry texts report the atomic masses of different elements with values from 4 to 7 significant figures. Sometimes in problems you will do later, it's helpful to round the atomic masses to the tenths place. For example:

Carbon (atomic mass 12.011) rounded to the tenths place becomes 12.0.

Hydrogen (atomic mass of 1.008) rounded to the tenths place becomes 1.0.

Aluminum (atomic mass 26.98) rounded to the tenths place becomes 27.0.

Sulfur (atomic mass of 32.06) rounded to the tenths place becomes 32.1.

Significant Figures in Multiplication and Division

When you multiply or divide measurements, your answer is limited by the number with the least number of significant figures. Suppose three students measure a box. Jane measures the length to be 3.12 cm, Carl measures the width to be 3.1 cm, and Donna measures the height to be 7.62 cm. To get the volume we multiply the three sides together.

$$3.12\,\text{cm} \times 3.1\,\text{cm} \times 7.62\,\text{cm} = 73.70064\,\text{cm}^3$$

However, we must round the calculated answer to 74 cm^3 because Carl made his measurement only to 2 significant figures.

When we calculate density, we divide mass by volume. That is

$$\text{Density} = \frac{\text{mass}}{\text{volume}}$$

Let's say we measure a sample's mass to be 7.4219 grams and its volume to be 1.23 mL. We divide 7.4219 by 1.23

$$\frac{7.4219\ \text{g}}{1.23\ \text{mL}}$$

the calculator answer is **6.034065041**

Because 1.23 mL has only 3 significant figures, we must round the answer to 6.03.

Significant figures are important, but don't get too bogged down with them. They will come naturally to you as you do more calculations.

When Do I Hit the EXP Button? Scientific Notation on the Calculator

Calculators make scientific notation easy. You just need to have the right calculator and know which buttons to push.

> Scientific calculators have a button that is marked [EE] (on most Texas Instruments calculators) or [EXP] (on most other brands). This is the button that lets you enter a number in scientific notation.

Let's take the number 7.45×10^8

To enter this number, we enter the coefficient 7.45, then press [EXP]. Two zeros appear at the right (smaller in size). These zeros are holding the places for the two digits of the exponent. (You will never need more than two.) We then enter the exponent 8. The display should look like this:

7.45 08

When you see this display, you read it as 7.45×10^8. Press [C] to clear the display and let's enter it again.

7.45 [EXP] 8 The display reads **7.45 08**

Enter 9.992×10^{12}

Press 9.992 [EXP] 12 The display reads **9.992 12**

A common mistake students make is to hit [×] before [EXP]. Don't do this! The "times 10 to the" is all included in [EXP].

So you do **not** enter

7.45 [×] [EXP] 8 or 7.45 [×] 10 [EXP] 8

you simply press 7.45 [EXP] 8

Let's enter 9.12×10^{12}

Press 9.12 [EXP] 12 The display reads **9.12 12**

Let's enter a number with a negative exponent: 7.216×10^{-9}

Enter 7.216 [EXP] 9 [+/-] or 7.216 [EXP] [+/-] 9 The display reads **7.216 -09**

[+/-] toggles a number between negative and positive. If you press [+/-] after [EE] or [EXP], it toggles the exponent between negative and positive. Press [+/-] a couple of times and see how the sign of the exponent changes back and forth.

[+/-] is **not** the same as [−]. [+/-] toggles between negative and positive. [−] is used only for subtraction.

Let's enter a few more numbers in scientific notation.

Enter the mass of an electron which is 1.6606×10^{-24} kilograms.

Press 1.6606 [EXP] 24 [+/-] The display reads **1.6606 -24**

Enter 10^5

When entering an exponential number where no coefficient is shown, you must supply the missing coefficient of 1. We can write 10^5 as 1×10^5.

Press 1 [EXP] 5 The display reads **1.0 05**

Enter 10^{-7}

Press 1 (EXP) 7 (+/-) The display reads **1.0 -07**

Multiplication and Division

If you multiply two numbers in scientific notation, just press (×) between the numbers. Let's multiply

$$(2.12\times10^{4})(7.89\times10^{-9}) =$$

Enter 2.12 (EXP) 4 (×) 7.89 (EXP) 9 (+/-) (=)

The display reads **1.67268 -04**
Since each number has three significant figures, we round the answer to 1.67×10^{-4}. Note the only time you enter (×) is between the two numbers.

Let's divide two numbers:

$$\frac{8.81\times10^{4}}{7.472\times10^{8}} =$$

8.81 (EXP) 4 (÷) 7.472 (EXP) 8 (=) The display reads **1.1790685 -04**

Since the least number of significant figures is 3, we round our answer to 1.18×10^{-4}. Let's try another:

$$\frac{10^{-14}}{10^{-9}} =$$

Remember to supply the missing coefficients of 1.

Press 1(EXP) 14 (+/-) (÷) 1 (EXP) 9 (+/-) (=)

The display reads **1.0000000 -05** which is written 1×10^{-5} or just 10^{-5}.

Now let's look at one last example that uses Avogadro's number (6.02×10^{23}), a number that you will see later in the class.

$$2.50\times10^{15}\times\frac{12.0}{6.02\times10^{23}} =$$

Press 2.5 (EXP) 15 (×) 12 (÷) 6.02 (EXP) 23 (=)

The display reads **4.9833887 -08**. Since each of the three numbers has three significant figures, we round the answer to 4.98×10^{-8}.

Switching Between Scientific Notation and Decimal

Sometimes you may want to convert an answer back and forth between scientific notation and decimal form. This will be most important to you on exams, especially multiple choice exams where the answer choices could be written in either form.

The answers to problems in the back of your textbook can also be given in either scientific notation or decimal form. Often students are confused when they look up a hard fought answer in the answer key and discover their answer looks wrong. Their answer may not be wrong. The textbook may have given the answer in scientific notation while the student calculated it in decimal form.

Texas Instruments Calculators

On Texas Instrument calculators we use [FLO] and [SCI] along with [2nd] to convert between scientific notation and decimal form. On TI-30X calculator, [FLO] and [SCI] are second functions written in blue above [4] and [5]. We use [2nd] to access these operations.

Pushing [2nd] [FLO] puts the calculator into normal decimal mode. (FLO comes from "floating point," computer programmers' jargon for a decimal number). In this mode, any number we enter will change to decimal form as soon as we hit another key (as long as the number will fit the display).

Pushing [2nd] [SCI] puts the calculator into scientific notation mode. Any number we enter will change to scientific notation as soon as we hit another key.

First let's change some numbers in scientific notation to decimal form.

Suppose you got 7.7×10^{-4} as an answer and want convert it to decimal form.

Enter 7.7 [EE] 4 [+/-]	The display reads	**7.7 -04**
Press [2nd] [FLO]	The display reads	**0.00077**

Now let's convert 8.449×10^{7} to decimal form.

Enter 8.449 [EE] 7	The display reads	**8.449 07**

Press (2nd) (FLO) The display reads 84490000

Now let's try to convert 6.419×10^{-17} to decimal form.

Enter 6.419 (EE) 17 (+/-) The display reads 6.419 -17

Now press (2nd) (FLO) The display still reads 6.419 -17

Nothing happens because a number in scientific notation with a –17 exponent has too many decimal places to fit into the 10 spaces on the calculator. A number in scientific notation must have an exponent of 9 or less, either plus or minus, to fit into the ten decimal places on a calculator. That is, a number must be between nine billion and nine-billionth.

Now let's take some decimal numbers and convert to scientific notation.

Suppose you got 0.000788 for an answer and want to change it to scientific notation.

Enter 0.000788 The display reads 0.000788

Press (2nd) (SCI) The display reads 7.88 -04

Convert 6,992,000 to scientific notation.

Enter 6,992,000 The display reads 6992000

Press (2nd) (SCI) The display reads 6.992 06

You can press (2nd) (FLO) to convert back to normal decimal form.

> If you press (2nd) (SCI) to convert an answer to scientific notation, the calculator remains in that mode until you press (2nd) (FLO) to put the calculator into normal decimal mode. If you enter a decimal number, it will change to scientific notation as soon as you press another function key such as (×), (÷) or (=).
>
> Likewise if you are in decimal mode when you enter a number in scientific notation, it will change to decimal form as soon as you hit another key (as long as the decimal form will fit in the display). In either of these cases the number is not affected, only the form of the number.

Let's look a few examples.

Press (2nd) (SCI) to put the calculator in scientific notation mode

Enter 0.0059 The display reads **0.0059**

Press (×) or (÷) The display changes to **5.9000000 -03**

Press (C) to clear the display. Now press (2nd) (FLO) to change your calculator to normal decimal mode.

Now enter 0.0059 The display reads **0.0059**

Now press (×) or (÷) The display still reads **0.0059**

Casio, Sharp and Most of the Rest

Most brands of calculators have a (MODE) button, usually in the upper left hand corner next to (INV). They also have a table printed under the readout display that looks something like this:

MODE	•	0	4	5	6	7	8	9
	SD	COMP	DEG	RAD	GRA	FIX	SCI	NORM

or this

MODE	0:COMP	4:DEG	5:RAD	6:GRA
	• SD	7:FIX	8:SCI	9:NORM

The two numbers you need in this table to convert between scientific notation and decimals are 8:SCI (scientific notation) and 9:NORM (for normal decimal form).

Pressing (MODE) (9) puts the calculator into normal decimal mode. In this mode, a number we enter will change to decimal form as soon as we hit another key (as long as the number will fit in the display).

Pressing (MODE) (8) (8) puts the calculator into scientific notation mode. Any number we enter will change to scientific notation as soon as we hit another key. The second number you enter after pressing (MODE) (8) tells the calculator how many places of the coefficient the calculator will display. For instance, if we press (MODE) (8) (4), the answer will be rounded to 4 digits. It's better to let the calculator display 8 digits, then you can choose the proper number of significant figures for an answer.

Let's change some numbers from scientific notation to normal decimal form.

Suppose you got 7.7×10^{-4} as an answer and want to change it to decimal form.

Enter 7.7 [EXP] 4 [+/-]	The display reads	7.7 -04
Press [MODE] [9]	The display reads	0.00077

Convert 2.568×10^{8} to decimal form.

Enter 2.568 [EXP] 8	The display reads	2.568 08
Press [MODE] [9]	The display reads	256800000

Now let's try to convert 7.489×10^{-13} to decimal form.

Enter 7.489 [EXP] 13 [+/-]	The display reads	7.489 -13
Now press [MODE] [9]	The display still reads	7.489 -13

Nothing happens because a scientific notation number with a –13 exponent has too many decimal places to fit into the 10 spaces on the calculator. A number in scientific notation must have an exponent of 9 or less, either plus or minus, to fit into the ten decimal places on a calculator. That is, a number must be between nine billion and nine-billionth.

Now let's convert decimal numbers to scientific notation.

Suppose you got 0.00489 as an answer and want to change it to scientific notation.

Enter 0.00489	The display reads	0.00489
Press [MODE] [8] [8]	The display reads	4.8900000 -03

Convert 2,090,000 to scientific notation.

Enter 2,090,000	The display reads	2090000
Press [MODE] [8] [8]	The display reads	2.090000 06

Press [MODE] [8] [8] to put the calculator in scientific notation mode.

If you press (MODE) (8) (8) to convert an answer to scientific notation, the calculator remains in that mode until you press (MODE) (9) to change to normal decimal mode. If you enter a decimal number, it will change to scientific notation as soon as you press another function key such as (×) or (÷).

Likewise, if you are in decimal mode when you enter a number in scientific notation, it will change to decimal form as soon as you press another key (as long as the decimal form will fit in the display). In either of these cases the number is not affected, only the form of the number.

Let's look at some examples.

Press (MODE) (8) (8), this puts the calculator in scientific notation mode:

Enter 0.0059 The display reads 0.0059

Press (×) or (÷) The display changes to 5.9000000 -03

Press (C) to clear the display. Now press (MODE) (9) to return your calculator to normal decimal mode.

Now enter 0.0059 The display reads 0.0059

Now press (×) or (÷) The display still reads 0.0059

Does the Answer Make Sense?

A calculator is a very handy instrument. It will make computations quick and easy. But it is only as good as the numbers you enter. Computer programmers have an acronym for this: GIGO. Garbage in, garbage out. If you are in a hurry on a test, it is easy to push the wrong button and not notice it. Let's look at some calculations and see if we can tell if the answer makes sense.

$$\frac{9.2\times10^8}{5.5\times10^5} = 1.7\times10^3$$

We can check our answer to see if it's in the right ballpark by looking at our exponents: $10^8 \div 10^5 = 10^3$. Also notice that 9.2 is a little less than twice 5.5, so the ratio of 1.7 is reasonable. Notice that the numerator is larger than the denominator so the answer has a positive exponent.

Let's look at another example:

$$\frac{7.9\times10^{-4}}{6.9\times10^{-9}} = 1.1\times10^{5}$$

We look at the exponents: $10^{-4} \div 10^{-9} = 10^{-4-(-9)} = 10^{-4+9} = 10^{5}$.

It checks out. In this case we divided a larger number by a smaller number and we get a number greater than one. Also notice that 7.9 is just a little larger than 6.9, so a ratio of 1.1 makes sense.

If we had hit the [×] button instead of the [÷] button by mistake we would have gotten the wrong answer 5.5×10^{-12}. By quickly checking the exponents we can spot this kind of error. The few seconds you spend checking your answers is time well spent.

> Remember that studies show that most students can raise their grades one or even two letters by doing nothing more than being neat and checking their work.

Look at the following problem and the **incorrect** answer given.

$$\frac{7.50\times10^{9}}{6.5\times10^{5} \times 3.5\times10^{3}} = 4.0\times10^{7}$$

Do a quick check of the exponents. Did your check tell you that the exponent in the answer should be around 10^{1} and not 10^{7}? ($10^{9} \div 10^{5} \div 10^{3} = 10^{(9-5-3)} = 10^{1}$)

Can you figure out what the student did wrong to arrive at the **incorrect** answer?

Doesn't it appear that the student carried out the calculation **incorrectly** as:
7.5 [EXP] 9 [÷] 6.5 [EXP] 5 [×] 3.5 [EXP] 3 [=]

Instead of the correct key sequence:

7.5 [EXP] 9 [÷] 6.5 [EXP] 5 [÷] 3.5 [EXP] 3 [=]

Remember that on page 30, we noted that all numbers in the denominator must be preceded by [÷] even if a × sign appears in front of the number. This student got the wrong answer because she forgot this rule.

> On multiple choice tests, the wrong answer you're likely to get is often one of the choices.

Let's look at one last example:

$$12 \times \frac{1}{1000} = 12{,}000 \ ?!$$

Here the student probably pressed [×] instead of [÷]. Since 1000 is in the denominator, it will make 12 one thousand times smaller instead of one thousand times larger.

Practice Problems *Answers on page 73*

Use your calculator. Round all answers to the appropriate number of significant figures.

1. $\frac{4.807 \times 1.85}{0.467 \times 5.00} =$

2. $\frac{96.8}{15.2 \times 0.0120} =$

3. $10.7 \times \frac{1.2}{3.8} \times \frac{14.2}{5.01} =$

4. $(4.16\times10^{5})(5.08\times10^{-8}) =$

5. $\frac{7.00\times10^{-6}}{5.871\times10^{4}} =$

6. $\frac{5.68\times10^{3}}{5.02\times10^{2}} =$

7. $(10^{5})(10^{-8})(10^{3}) =$

8. $\frac{10^{-14}}{10^{-9}} =$

9. $3.15\times10^{18} \times \frac{342}{6.02\times10^{23}} =$

10. $1.00\times10^{15} \times \frac{1.66\times10^{-24}}{12.0} =$

Study Aid 3 How to Take Notes in Class

Sit in the front and middle of the class, especially in large and noisy lecture halls. It will be easier to hear and see. It will also force you to stay alert and take notes because you will be right in the line of sight of the instructor!

Get a large notebook with pockets or a looseleaf notebook so you will have a place to put the syllabus and any other handouts. Don't be afraid to use paper. One of the worst things you can do is to write your notes small and all scrunched together. You will go blind trying to decipher them later, and you will have no room to add extra information or to clarify something.

Prepare for the lecture before class by:

- reviewing the notes from the previous class.
- reviewing the homework.
- glancing over the chapter sections that will be covered that day. Don't worry if you don't understand everything. At least the instructor's terms will be familiar.

During class:

- Write on the right hand two-thirds of the page. Save the left hand side for notes when you review after class and for writing key words that will help you when studying for a quiz or test. (Study skills texts refer to this method as the Cornell Method of notetaking.)
- Write the date at the beginning of each day's notes.
- Take notes as completely as you can while still understanding.
- Write down each step of every problem even if you do not understand the step. You can always ask about it later.
- Write a question mark next to anything you don't understand. Come back to it later. Try to find the topic in the text or get an explanation from the professor, a tutor, or better yet, from a classmate.
- Don't just listen. Be an ACTIVE LEARNER. Nod when you understand something. Give yourself (and the professor) some positive reinforcement!
- Listen for clues such as, "This definition is very important," or "I don't have time to cover this in class, but it is required," or "This is not covered in your textbook."

- Use shorthand and abbreviations so that you can write quickly but understandably. Examples are BP for boiling point and SH for specific heat. The first time you use an abbreviation write its translation, then continue with just the abbreviation.
- Don't get lost in the five minute rustle. Five or ten minutes before class ends, some students start rustling their books, papers and bookbags in anticipation of running off to their next exciting activity. This is especially a problem in large lecture halls. Don't neglect these last minutes of class; you can miss some important information. Being able to hear clearly during the five minute rustle is also a good reason for sitting in the front and center.

As soon as possible after class:

- Review your notes while they are still fresh in your mind. This will help to move the information from your short-term memory to your long-term memory. (If at all possible, never, never schedule a class immediately after any course that you feel is going to be difficult for you. You need the hour immediately after that class to go over notes and begin homework while the material is still fresh. Studies tell us that if you go over material within an hour or so after class, you will remember about 80% of what you have learned. If you wait until later, you will forget 80% of what you have learned. That's a big difference!)
- Mark your notes. Highlight with a marker. Write notes in the left-hand margin. Take ownership of your notes and your text.
- Above all else never, never miss a lecture! Chemistry is cumulative. What is presented tomorrow will depend upon your knowledge of what was covered today. If for any reason you have no choice other than to miss a class, arrange to get someone's notes and go over them before the next lecture. (Be sure to get notes from someone who is as good a notetaker as you are!)

CHAPTER 4

Rulers, Weights, and Measuring Cups Measurement and the Metric System

Chemistry is a quantitative science. You can look at the world two ways: qualitatively or quantitatively. The redness of an apple, the stickiness of an adhesive, the smell of a perfume, or even your personality are qualitative properties. Your height, your weight, your bank balance, or the amount of gasoline you buy are quantities. Sometimes there is a fine line between qualities and quantities. The redness of an apple can be measured as a wavelength of so many nanometers of light. The stickiness of an adhesive can be measured in the pounds of force it takes to peel it off. And psychologists even try to quantify something as indefinable as a personality with various psychological tests. On the other hand, for some people, even their bank balance is qualitative. Instead of an exact balance they say, "I've got some," or "I've got enough to cover that check."

You may be a little dismayed that chemistry often deals with quantities. But remember, you work with quantities constantly in your everyday life. You buy 10 gallons of gasoline. You add 1/2 teaspoon of salt to a recipe for bread. You weigh yourself in pounds on your bathroom scale.

Gallons, teaspoons, and pounds are examples of units that you use in everyday life. Units are standards of measurement we all agree upon so that we talk the same language. If the author of a recipe tells you to add 2 teaspoons of salt, you, she, and the teaspoon manufacturer have to agree on the definition of a teaspoon. If there were no agreement on units like teaspoons, it would be difficult to trade recipes!

Likewise, if we all did not agree on temperature, it would be hard to tell someone else how cold it is outside. What if the weatherman, instead telling us a temperature, just reported it was "hot," "cool," "cold," or "real, real cold"? What might be "cool" to him or her might be "real, real cold" to you. Fortunately most weather reporters give us the temperature as a quantity with units of degrees. If the weatherman says the temperature is 5°F, you can make your own judgment based on your own experience of 5°F temperature.

How Big are the King's Feet? Measurement and the English System

Before we go into the metric system units you will use in chemistry, let's look at our own English units that we are more familiar with. In any system of measurement, there are three fundamental properties we measure: length, mass, and volume.

Length is the measurement of distance. We measure large distances, like the distance across town or the distance between cities, in miles. We measure medium size lengths such as the size of buildings and rooms in feet and yards, And we measure small distances, like the size of postcards, in inches.

As you may know, these units are very ancient. A "foot" was decreed to be the size of the King's foot. (The king that gave us our modern foot must have had quite big feet!) An inch was the size of a knuckle. Henry I of England decreed that a yard was to be the distance from his nose to his thumb when his arm was outstretched.

The next fundamental unit is mass. Mass and weight are often confused with each other in everyday language. Mass is an object's resistance to change in motion. A semi-truck has much more mass than a little Volkswagen. It's hard to get a semi going when it's stopped (as we all have experienced as we've sat behind one after the light has turned green). Similarly it is very hard to stop a semi once it's going fast.

Weight, on the other hand, is the gravitational attraction for a mass. Thus if you weigh 180 pounds on earth, you will weigh 30 pounds on the moon, which has 1/6 the gravity of earth. And in outer space you will be weightless. You will be weightless but not massless. If you get clanked on the head in outer space by a weightless hammer, you would still feel it. It is the mass, not the weight, that you feel.

The pound is actually a weight unit, not a mass unit. However, in common usage, we do not make a distinction between weight and mass. The distinction is not important unless we are trying to use a spring scale, which depends on gravity, or we are on the moon or in the space shuttle.

The last fundamental property is volume. Volume is the measurement of the amount of space something occupies. Your body takes up space. You can see this the next time you get in the bathtub and watch the water level as your body displaces some of the water.

We have many English units for volume: gallons, quarts, pints, cups, tablespoons, teaspoons, fluid ounces. We buy milk and ice cream in gallons, quarts, and pints. We measure recipes in cups, tablespoons, and teaspoons and buy medicine by the fluid ounce.

Another way to think of volume is length cubed. We live in three-dimensional space: an object has length, width, and depth. If we have a box that is 1 foot high, 1 foot wide, and 1 foot deep, the box has a volume of 1 foot^3, which we read as one cubic foot. An aquarium that is 2 feet wide, 1 foot high and 1 foot deep has a volume of 2 feet^3. Volumes of refrigerators and swimming pools are measured in cubic feet. If you use natural gas, look on your bill and you will see that gas usage is reported in cubic feet.

Oh, Those French—The Metric System

The problem with our English system, as you might already be aware, is that it is very confusing to convert between units. Imagine if you came here from a foreign country and had to memorize 5,280 feet to a mile, 4 quarts to a gallon, 16 ounces to a pound, 2000 pounds to a ton, etc. You would go crazy!

That is just what some French revolutionaries thought in the 1790s, during the French Revolution. They developed a system of measurement we now call the "metric system." Their idea was to base their new system of units on powers of ten, just like our number system. They reasoned that it would be easy to use and remember. They did not want to base length on a king's arms or feet, since they had overthrown the King, so they defined the metric unit of length, the meter, as 1/10,000,000 of the distance from the equator to the north pole (on the line that goes through Paris, of course). This length worked out to be a little longer than a yard.

They went on and did the same thing with mass, volume and temperature units, as we shall soon see.

Well, the ball started rolling in France, and now every country in the world, including England, is metric with one exception—the USA! Actually our government officially adopted the metric system several times, first in 1866 and then again in 1974 and 1985. But except for scientific use, the metric system never caught on. If it had, and we were raised with the metric system, you would not need to read this chapter.

So, the problem most Americans have with the metric system is that it is not familiar. It does not come naturally to us like our inches, quarts, and pounds that we know so well. What we want to do in the rest of this chapter is to help make the metric system familiar for you. If working metric problems is not just manipulating numbers (as it is for most people in beginning chemistry classes), you will make far fewer mistakes. It will also be easier to learn, and you will remember it much longer.

Give Them a Centimeter and They'll Take a Kilometer

The fundamental metric unit of length is the meter, which is a little longer than a yard (39.37 inches to be exact). For measuring small distances we need to have subdivisions of the meter. In the English system, an inch is a subdivision of the foot. Subdivisions are handled elegantly in the metric system with prefixes. The neat thing is that the prefixes will be the same no matter what units we attach to them. Once you learn them for one unit, they work the same way for any other unit. Table 4.1 lists the metric prefixes. Let's see how the prefixes work with the meter.

- **deci**—the prefix deci means $\frac{1}{10}$ or 0.1. The Latin root "deci" is where we get our base 10 "decimal" system.

If we put the prefix deci in front of meter we get

$$1 \text{ decimeter} = 0.1 \text{ meter}$$

If a decimeter is 0.1 (one tenth) of a meter, then we can also say there are 10 decimeters (dm) in a meter.

$$10 \text{ dm} = 1 \text{ m}$$

Figure 4.1 shows these two relationships.

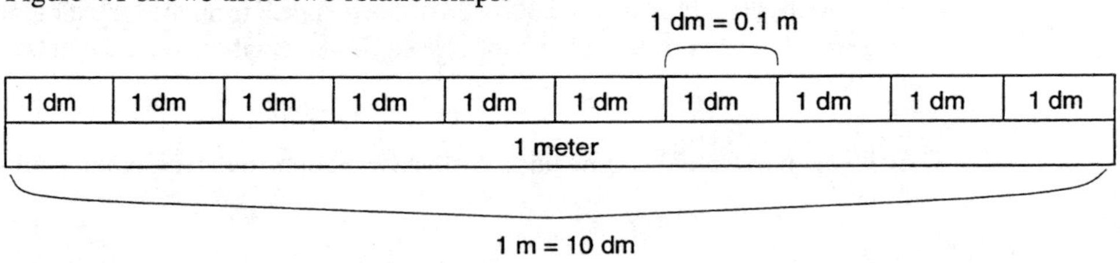

Figure 4.1

A decimeter is about the length of a finger.

- **centi**—the prefix centi means $\frac{1}{100}$ or 0.01.

"centi" comes from the same Latin root as cent which is 0.01 of a dollar.

If we put the prefix centi in front of meter we get

$$1 \text{ centimeter} = 0.01 \text{ meter}$$

If a centimeter is 0.01 (one hundredth) of a meter, then we can also say there are 100 centimeters in a meter.

$$100 \text{ cm} = 1 \text{ m}$$

A centimeter is about the width of a fingernail.

- **milli**—the prefix milli means $\frac{1}{1000}$ or 0.001

"milli" comes from the same root as *mill* in property taxes. A mill is 0.001 of a dollar.

If we put the prefix milli (abbreviated m) in front of meter we get

$$1 \text{ mm} = 0.001 \text{ meter}$$

If a millimeter is 0.001 (one thousandth) of a meter, we can also say there are 1000 millimeters (mm) in one meter.

$$1000 \text{ mm} = 1 \text{ m}$$

A millimeter is about the thickness of a dime.

- **micro**—the prefix micro means one millionth, which we can write as

$$\frac{1}{1{,}000{,}000} \text{ or } \frac{1}{10^6} \text{ or } 10^{-6}.$$

When the numbers get this small, you can see it is easier to put them in exponential form.

If we put the prefix micro (abbreviated μ) in front of meter we get

$$1\,\mu\text{m} = 10^{-6} \text{ meter}$$

or, written with a positive exponent

$$10^6\,\mu\text{m} = 1\text{m}$$

Since a micrometer is one millionth (10^{-6}) of a meter, there are a million micrometers (10^6) in a meter. The cells of your body are about 3 to 5 μm in diameter.

nano—nano means one billionth, which we can write as $\frac{1}{1{,}000{,}000{,}000}$ or $\frac{1}{10^9}$ or 10^{-9}. So if we put the prefix nano (abbreviated n) we get:

$$1 \text{ nanometer} = 10^{-9} \text{ meter}$$

or in whole numbers

$$10^9 \text{ nm} = 1\text{m}$$

A nanometer is a billionth of a meter, therefore there are a billion nanometers in a meter. A nanometer is about the size of a molecule in your body.

Table 4.1: The Most common Metric Prefixes

Prefix	Symbol	Base Unit Multiplied By
kilo	k	1000 or 10^{3}
deci	d	0.1 or 10^{-1}
centi	c	0.01 or 10^{-2}
milli	m	0.001 or 10^{-3}
micro	μ	0.000001 or 10^{-6}
nano	n	0.000000001 or 10^{-9}

Keep in mind with all these prefixes we are taking a meter and cutting it into smaller and smaller pieces: 10 pieces for the decimeter, 100 pieces for the centimeter, 1,000 pieces for the millimeter, 1,000,000 pieces for the micrometer, and 1,000,000,000 pieces for the nanometer.

Besides subdivisions, we can also have multiples of the basic unit. In the English system the mile is a multiple of the foot (1 mile = 5,280 feet). In chemistry the only multiple we commonly use is kilo.

$$\text{kilo} = 1000$$

So

$$1 \text{ kilometer} = 1000 \text{ meters}$$

If you lay 1,000 meter sticks end-to-end, you get a kilometer, which is just a little over six tenths of a mile. To get a feeling for how long a kilometer is, the next time you drive from your house, notice when the odometer moves 0.6 miles. That distance is about a kilometer.

Students often mix up millimeters and kilometers when they do problems. If you make the metric system concrete, this will not happen to you. A millimeter is the thickness of a dime. If you stacked 1000 dimes, you would get about a meter (1000 mm = 1 m) If you lay a thousand meter sticks (each a little longer than a yard) end to end, you get 1 kilometer (1 km = 1000 m).

A Gram of Prevention is Worth a Kilogram of Cure

The same prefixes we used for the meter can be used for the gram. However, in chemistry decigrams and centigrams are rarely used. But we often use:

$$1 \text{ milligram} = 0.001 \text{ gram}$$

or in whole number form

$$1000 \text{ mg} = 1 \text{ g}$$

$$1 \text{ microgram} = 10^{-6} \text{ gram}$$

or written with a positive exponent

$$10^{6}\mu\text{g} = 1 \text{ g}$$

$$1 \text{ nanogram} = 10^{-9} \text{ gram}$$

or written in whole number form

$$10^{9} \text{ ng} = 1\text{g}$$

A gram is about 1/28 of an ounce or about the weight of an aspirin pill. A milligram is about the weight of a gnat or a pinch of salt.

We can also use the prefix kilo with the gram

$$1 \text{ kg} = 1000 \text{ g}$$

A kilogram is about two pounds (2.2 pounds to be exact).

Just a Milliliter of Sugar Helps the Medicine Go Down

The metric unit of volume is the liter. The liter is defined as a cube that is 10 cm on each side. It is 10 cm wide, 10 cm high and 10 cm deep. We multiply

$$10 \text{ cm} \times 10 \text{ cm} \times 10 \text{ cm} = 10^{3}\ 10^{3} = 1000 \text{ cm}^{3}$$

$$10 \times 10 \times 10 \times \text{cm}^{3} \times \text{cm}^{3} \times \text{cm}^{3} = 10^{3} \text{ cm}^{3} = 1000 \text{ cm}^{3}$$

Units multiply and divide just as numbers do. So $10 \times 10 \times 10 = 10^{3}$ and $\text{cm} \times \text{cm} \times \text{cm} = \text{cm}^{3}$. The big one liter cube has 1000 little centimeter cubes in it.

If you fill a one liter cube with water and then pour the water into a bottle, you still have one liter. The shape of the liter does not matter. We defined a liter as a cube, but now we can forget about the cube and just remember the conversion:

$$1 \text{ Liter} = 1000 \text{ cm}^{3}$$

A liter is just a little bigger than a quart. A liter bottle of soda gives you about 6% more than a quart.

We can use the metric prefixes to make subdivisions of the liter.

$$1 \text{ deciliter} = 0.1 \text{ liter}$$

or using whole numbers

$$10 \text{ dL} = 1 \text{ L}$$

Deciliters are not used very often in chemistry, but they are often used in hospitals where blood test results are reported in milligrams per deciliter. A deciliter is about a half cup.

The primary subdivision of the liter we use in chemistry is the milliliter.

$$1 \text{ milliliter} = 0.001 \text{ liter}$$

or using whole numbers

$$1 \text{ L} = 1000 \text{ mL}$$

Since we also know that

$$1 \text{ L} = 1000 \text{ cm}^3$$

we conclude

$$1 \text{ mL} = 1 \text{ cm}^3$$

A milliliter is exactly the same size as a cubic centimeter. This information will be important later in density problems where milliliters and cubic centimeters are used interchangeably. In hospitals, cubic centimeters are often called "cc's". So we can say

$$1 \text{ mL} = 1 \text{ cm}^3 = 1 \text{ cc}$$

A sugar cube about the width of your fingernail is about the size of a milliliter. A teaspoon is about 5 milliliters.

Other Metric Volumes

You may have noticed we did not use the metric prefix kilo with liters. It is never used. Instead for volumes larger than the liter the cubic meter is used. A cubic meter is a cube that is 1 meter on each side. Imagine a box that is a little longer than a yard on each side. A cubic meter holds 1000 liters.

$$1 \text{ m}^3 = 1000 \text{ L}$$

So a cubic meter is really a "kiloliter." We do not use the term "kiloliter" because the meter is considered a more fundamental unit than the liter. You use cubic meters for large volumes like swimming pools or large tanks. Many natural gas companies now use cubic meters instead of cubic feet for billing.

Sometimes You Have to Sigh
A Note on SI Units

In 1960 scientists revised and further standardized the metric system to give it more uniformity. They called the revised system The International System of Units or SI (after the French Systéme Internationale). The basic SI unit of mass is the kilogram, rather than the gram as it is in the metric system. Likewise the SI unit of volume is the cubic meter (m^3) rather than the liter. The meter is the basic unit of length in both systems.

In most cases you will use metric units. But occasionally you may have to use SI units. When you study the properties of gases you will use the SI unit of temperature, the Kelvin (K), rather than the metric unit of degrees, Celsius ($^\circ C$). And when you study heat, you will learn both the SI unit, the Joule (J), and the metric unit, the calorie (cal).

Reading the Want Ads
Kilodollars and other Units

The metric prefixes can be used with any units. If you read the want ads, you might read salary "28-35K$"or simply "28-35K." These ads are using "kilodollars."

1 kilodollar = 1000 dollars

So 25 to 35 kilodollars is $28,000- $35,000.

It doesn't matter if you know what the unit means. For instance, you know from their prefixes that

1 kilojoule = 1,000 joules

100 centipoise = 1 poise

Table 4.2 lists real things the size of each metric unit. Study this table and try to think of your own examples. After a while you will begin to make metric measurements concrete and real so that you will not just be memorizing numbers. Then you will really start thinking like a European and making conversions will become a simple matter. Table 4.3 lists in one place all the metric conversions you will need to memorize. The next chapter will show you a very easy method to convert between metric units and to convert metric units to English units.

Table 4.2: Metric Sizes of Everyday Objects

Length		Mass	
meter	a yardstick, about waist high	gram	two aspirin tablets
decimeter	length of finger	milligram	a gnat
centimeter	width of fingernail	kilogram	a little over 2 lbs of cheese
millimeter	thickness of a dime	**Volume**	
kilometer	a little over half a mile	liter	a quart
		deciliter	half of a cup
		milliliter	1/5 of a teaspoon

Table 4.3: The Most Common Metric Conversions

Length	Mass	Volume
1m = 10 dm	1 g = 1000 mg = 10^3 mg	1 L = 10 dL
1 m = 100 cm	1 g = 10^6 μg	1 L = 1000 mL = 10^3 mL
1 m = 1000 mm = 10^3 mm	1 g = 10^9 ng	1 mL = 1 cm^3 = 1cc
1 m = 10^6 μm	1 kg = 1000 g = 10^3 g	1 m^3 = 1000 L = $10^6 cm^3$
1 m = 10^9 nm		
1 km = 1000 m = 10^3 m		

Study Aid 4 Learning to Relax

It's hard to study effectively and you certainly won't perform your best on an exam if your body is overly tense. Also when you are tense, things always look worse than they are. However, some stress is normal when you enter into the unknown. And a little tension keeps you from getting too complacent and making careless errors. It is the excess tension that you need to be able to control.

There are three factors you can consciously control to reduce your stress and increase the effectiveness of your studying: posture, muscle tension and breathing. We will discuss just a few techniques you can try with each factor. There are many books available on relaxation; we encourage you to seek out these resources.

Posture

When we are concentrating hard, especially when we are tired, we often tend to pull our neck down into our shoulder blades, collapse our chest and let the lower spine bow out. Sitting with the body in this position tends to make us tired and tense. Our muscles work overtime to hold the body up and the lungs cannot breathe efficiently.

When you are studying, and especially when you are taking a test, imagine a string is tied to the crown of your head lifting it gently away from your shoulder blades and elongating your neck and spine. When you are sitting or standing, imagine your spine gently pulled upward. Think tall but not rigid. When you sit at your full height, it will help you will feel at ease and confident.

Muscle Tension

A tense muscle uses more energy and requires more oxygen and nutrients than a muscle at rest. But a tense muscle constricts the capillaries that deliver oxygen and nutrients to it, which leads to fatigue of the muscle. The exhausted muscle becomes even more tense starting a downward cycle. There are three things you can do to relieve muscle tension: stretching, massage and movement.

While you are studying, remember to stretch. Stretch your hands above your head. Clasp your hands behind your back and expand your chest. Stand up periodically and stretch any way that feels good. If you are at home, lie on the floor and stretch your arms over your head.

Gently massage your face and scalp. Then massage your neck and shoulders. Massaging your feet will refresh you during a long study session.

Take a walk. Walking is one of the best ways to relax. Even a short walk around the block will help you relax and feel more alert.

Periodically look out the window or across the room to relax your eye muscles. You can also rub your hands together to create a gentle friction and then place your the hollows of your palms over your eyes. The warmth will relax your eye muscles.

Breathing

Your breath is tied to your emotions. Have you noticed that after a traffic skirmish your breath becomes rapid and shallow. This is one symptom of what is known as "fight or flight," a remnant of the days when we had to either confront a danger or run. In the modern world this reaction does little to help us. The next time a traffic incident occurs, take a long slow breath and feel what happens. You should feel more calm and in control.

The same "fight or flight" reaction can happen on a smaller scale when you are studying unfamiliar material or when you are taking an exam. Here are some breathing techniques you can try anytime you feel tense.

Breathe in slowly through your nose as you count to five. Then breathe out slowly. Pause for a second when you have completely exhaled, then repeat for a few more breaths. When you inhale, expand your abdomen first to fill the lower lungs. Then let your chest rise to fill the upper lungs. Don't force the breath; let it come easy and natural. You can do this exercise a few minutes before you take an exam.

You can write "breathe" on the top of your exam to remind you to take breathe deeply while taking a test.

CHAPTER 5

It's the "One" Thing To Do Unit Conversions

In the last chapter you looked at conversions between metric units. Those conversions are listed for you in one convenient place in Table 4.3. You also started to make the metric units more familiar by relating their sizes to everyday items and distances.

In this chapter we will show you a very easy way to convert between metric units and also between metric and English units. We call the process "unit conversions." It goes by other names in some texts such as "factor labeling," "unit analysis," or "dimensional analysis." The process is the same no matter what we call it.

Most of the math you will do on your first test will involve unit conversions. This is a powerful technique to learn. It is not just memorizing formulas. It is a way of thinking that you can use on your first test, on problems later in the class, and even to solve problems in everyday life and your career.

What Do Unit Conversions Mean to Me?
Doing Conversions

Before we look at the metric system, let's start with a English relationship we know.

12 inches = 1 foot

This statement is an equality, that is, we are saying 12 inches is exactly the same as 1 foot. If we multiply or divide both sides of an equality by the same thing, we do not change the relationship. It remains an equality. Let's divide both sides by 1 foot.

$$\frac{12 \text{ inches}}{1 \text{ foot}} = \frac{1 \cancel{\text{foot}}}{1 \cancel{\text{foot}}} = 1$$

Anything divided by itself equals one and units cancel just as numbers do. Therefore we have derived a fraction, $\frac{12 \text{ inch}}{1 \text{ foot}}$, that is equal to one.

Now let's divide the equality by 12 inches:

$$\frac{\cancel{12} \cancel{\text{inches}}}{\cancel{12} \cancel{\text{inches}}} = \frac{1 \text{ foot}}{12 \text{ inches}} = 1$$

Therefore from the equality

$$12 \text{ inch} = 1 \text{ foot}$$

we can derive two fractions that are equal to one.

$$\frac{12 \text{ inches}}{1 \text{ foot}} \text{ and } \frac{1 \text{ foot}}{12 \text{ inches}}$$

Notice that one conversion is the reciprocal of the other, that is, it is the other turned upside down.

Now let's use these conversion factors to do some conversions between inches and feet. These conversions may seem very simple, but you will do exactly the same process when you convert between units in the metric system and when you convert between moles and grams later in the class.

Remember to look for patterns!

Let's convert 4.5 feet to inches.

$$4.5 \cancel{\text{feet}} \times \frac{12 \text{ inch}}{1 \cancel{\text{foot}}} = 54 \text{ inch}$$

We multiplied by the conversion factor with the foot units on the bottom so they cancel, and we end up with the unit we are trying to solve for, inches. No matter what the units are, the pattern for converting will be the same:

$$\cancel{\text{old units}} \times \frac{\text{new units}}{\cancel{\text{old units}}} = \text{new units}$$

Now let's convert 30 inches to feet.

$$30 \text{ inch} \times \frac{1 \text{ foot}}{12 \text{ inch}} = 2.5 \text{ feet}$$

The old units, inches, cancel and we end up with the new units, feet.

Always label the numbers in a problem with units. Without units many mistakes can go undetected!

One of the great things about unit conversions is they are self-checking. What if we made this mistake:

$$30 \text{ inch} \times \frac{12 \text{ inch}}{1 \text{ foot}} = ?!$$

You can immediately see that our units do not cancel. So you know right away that the conversion needs to be inverted. Using unit conversions makes it almost impossible to make a mistake.

But Where Do You Find Conversions?

You have used some conversions so often in your life you just know them. These are conversions like 12 inches to 1 foot and other English system conversions such as 16 fluid ounces to 1 pint or 4 quarts to 1 gallon. You may need to look up other conversions. For a while you will need to look up the metric conversions in Table 4.3. We hope that these conversions will also become second nature to you soon. But for now, don't be ashamed to look them up.

Now let's try some metric conversions. Let's convert 2.57 meters to millimeters. Before we do the conversion, look up the conversion between millimeters and meters in Table 4.3. It is

$$1000 \text{ mm} = 1 \text{ m}$$

You can write the two conversions we get from this equality:

$$\frac{1000 \text{ mm}}{1 \text{ m}} \text{ or } \frac{1 \text{ m}}{1000 \text{ mm}}$$

Now let's do the conversion. First we write down the given quantity, then multiply by a conversion factor.

$$2.57 \text{ m} \times \frac{1000 \text{ mm}}{1 \text{ m}} = 2570 \text{ mm}$$

We chose the first conversion with meters in the denominator so that they cancel the meter units in the given quantity. On the calculator:

Press 2.57 [×] 1000 [=] The display reads **2570**

Notice that since the 1000 is in the numerator, we multiply by 1000 and the decimal point moves three places to the right, that is, it gets larger by 1000.

Now let's convert 435 mm to m.

We use the same conversion factor 1000 mm = 1 m

$$435\ \cancel{\text{mm}} \times \frac{1\ \text{m}}{1000\ \cancel{\text{mm}}} = 0.435\ \text{m}$$

Since the 1000 is in the denominator, we divide. On the calculator:

Press 435 [÷] 1000 [=] The display reads **0.435**

If we use the wrong conversion, as shown below, the units will not cancel and we will see that we have to turn the conversion upside down.

$$435\ \text{mm} \times \frac{1000\ \text{mm}}{1\ \text{m}} = \ ?!$$

Beware! There is one way you can make a mistake that will be hard to catch using unit conversions. That is if you memorized the metric conversion between meters and millimeters **incorrectly** as

$$1\ \text{mm} = 1000\ \text{m} \quad \text{(incorrect!)}$$

If you did this, you would set up the problem in a way which looked correct to you, because the units would cancel. But it would be wrong because you were using the incorrect numbers in the conversion. Let's show what would happen if you used the incorrect metric conversion to convert 1.662 meters to millimeters:

$$1.662\ \cancel{\text{m}} \times \frac{1\ \text{mm}}{1000\ \cancel{\text{m}}} = 0.001662\ \text{mm ?!!} \quad \text{(incorrect!)}$$

Look carefully at the incorrect answer. Does it make sense? If you have made the metric system concrete, you will recall that a millimeter is only about the thickness of a dime, while a meter is about a yard. So your answer is saying that something that is about 1¾ yards long is about 1/1000 the thickness of a dime. Of course this comparison could not be correct! Now look at the conversion you wrote and you will realize that you are saying the thickness of a dime is the same as a thousand meter sticks! You then catch your mistake and write the conversion correctly as $\frac{1000\text{ mm}}{1\text{ m}}$ and redo the problem:

$$1.662\ \cancel{\text{m}} \times \frac{1000\text{ mm}}{1\ \cancel{\text{m}}} = 1662\text{ mm}$$

Remember, to a person from France, 1 mm equals 1000 m would be as immediately absurd as 12 feet equals 1 inch is to you.

How Do I Move the Decimal Point? Metric Conversions

Now let's do a few more metric conversions. Again, all the conversions are found in Table 4.3.

Convert 27 cm to m.

$$27\ \cancel{\text{cm}} \times \frac{1\text{ m}}{100\ \cancel{\text{cm}}} = 0.27\text{ cm}$$

Convert 4.21 kg to grams.

$$4.21\ \cancel{\text{kg}} \times \frac{1000\text{ g}}{1\ \cancel{\text{kg}}} = 4210\text{ g}$$

Convert 35.0 mL to liters.

$$35.0\ \cancel{\text{mL}} \times \frac{1\text{ L}}{1000\ \cancel{\text{mL}}} = 0.0350\text{ L}$$

Note on Significant Figures

Exact numbers, such as those in conversion factors, do not affect the number of significant figures in a calculation. There are exactly 100 centimeters in one meter. There are exactly 1000 grams in one kilogram and 1000 milliliters in one liter. So in the three examples above, the answers have the same number of significant figures as the starting quantities. Likewise in the English system there are exactly 12 inches to a foot, 4 quarts to a gallon, 16 ounces to a pound, etc.

Multiple Conversions

Sometimes you may have to do more than one conversion in a problem. This often happens when neither of the units is the basic unit.

Convert 0.25 km to cm.

First we convert km to meters, the basic unit, then we convert meters to cm.

That is, we go km ⇨ m ⇨ cm.

$$0.25\ \cancel{\text{km}} \times \frac{1000\ \cancel{\text{m}}}{1\ \cancel{\text{km}}} \times \frac{100\ \text{cm}}{1\ \cancel{\text{m}}} = 25000\ \text{cm}$$

Or in scientific notation: 2.5×10^4 cm

Now the nice thing about using the calculator is that you can line up all the conversions in one problem. On your calculator:

Enter 0.25 [×] 1,000 [×] 100 [=] The display reads **25000**

You can then press [MODE] [8] [8] to convert to scientific notation if you choose. The answer then becomes 2.5×10^4 cm.

Then press [MODE] [9] to convert back to decimal form. Otherwise your calculator will remain in scientific notation—as we warned you in Chapter 3.

Let's try another. Convert 7.44×10^4 cm to kilometers.

The conversion will go like this: cm ⇨ m ⇨ km.

$$7.44 \times 10^4\ \cancel{\text{cm}} \times \frac{1\ \cancel{\text{m}}}{100\ \cancel{\text{cm}}} \times \frac{1\ \text{km}}{1000\ \cancel{\text{m}}} = 0.744\text{km}$$

Press 7.44 [EE] 4 [÷] 100 [÷] 1000 [=] The display reads **0.744**

Press [MODE] [8] [8] to convert to scientific notation 7.44×10^{-1}

Switching Systems
Metric-English Conversions

What if you want to convert between the English system and the metric system? Table 5.1 lists some common conversions between the two systems. Your textbook will have a similar table. Your instructor may want you to memorize a few conversions, such as 1.61 km to 1 mile or 1.06 qt to 1 L, but you probably won't be asked to memorize them all. However, you do need to know where to find them and how to use them.

Table 5.1 Metric-English Conversions

Length	Mass	Volume
1 m = 39.4 in.	1 kg = 2.20 lbs.	1 L = 1.06 qt
2.54 cm = 1 in.	454g = 1 lb.	1 mL = 5 tsp
1.61 km = 1 mile		946 mL = 1 L

These conversions work the same as any other unit conversions. Let's try some.

How many miles is a 10.0 kilometer race (often called simply a "10 K")?

We look up the conversion between miles and kilometers in Table 5.1 and find

$$1.61 \text{ km} = 1 \text{ mile}$$

So lining up the conversions:

$$10.0\,\cancel{\text{km}} \times \frac{1 \text{ mile}}{1.61\,\cancel{\text{km}}} = 6.21 \text{ miles}$$

We round the answer off to 3 significant figures.

A piece of cloth is 47 inches long. How many centimeters long is it?

Look up the conversion between cm and inches in Table 5.1. We find it is

$$2.54 \text{ cm} = 1 \text{ in.}$$

So can write the conversion so that the inch units cancel

$$47\,\cancel{\text{inch}} \times \frac{2.54 \text{ cm}}{1\,\cancel{\text{inch}}} = 120 \text{ cm}$$

Press 47 [×] 2.54 [=] The display reads **119.38**

We round the answer off to 2 significant figures to 120 cm, or in scientific notation to clearly show that we have two significant figures 1.2×10^2 cm.

What if you want to order 1.5 lbs of cheese in Canada where they use the metric system? How many kilograms do you order? Look up the conversion.

$$1.5\ \cancel{\text{lb}} \times \frac{1\ \text{kg}}{2.2\ \cancel{\text{lb}}} = 0.68\ \text{kg}$$

Sometimes we need to do more than one conversion depending on what conversions we find in the table. For example, suppose we want to convert 0.500 gallons to liters. When we look up the conversion in Table 5.1, we only find

$$1.06\ \text{qt} = 1\ \text{L}$$

Even though we do not find a conversion between gallons and liters, we can still do the problem because we know the English relationship

$$4\ \text{qts} = 1\ \text{gal}$$

So we can use gallons ⇨ quarts ⇨ liters.

$$0.500\ \cancel{\text{gal}} \times \frac{4\ \cancel{\text{qt}}}{1\ \cancel{\text{gal}}} \times \frac{1\ \text{L}}{1.06\ \cancel{\text{qt}}} = 1.89\ \text{L}$$

Press .5 [×] 4 [÷] 1.06 [=]

The display reads **1.886792443**

We round the answer off to 3 significant figures.

Reading the Tables
Where Do You find Conversions?

There are three ways you can find conversions. First, conversions are often listed in tables like the metric conversions in Table 4.3 or the metric-English conversions in Table 5.1 in this book. You will find similar tables in your textbook. Second, there are many conversions you know because you have worked with them so often in everyday life. You are probably thoroughly acquainted with English conversions such as:

5,280 feet = 1 mile
100 cents = 1 dollar
2 pints = 1 qt
4 quarts = 1 gallon
16 ounces = 1 pound

A third source of conversions is to figure them for yourself. Anytime you see the word "per" you have a conversion because "per" means to divide. For example, if a typist types 75 words per minute, you can write this conversion: $\frac{75 \text{ words}}{1 \text{ min}}$. If you earn \$8.00 per hour, you earn \$8.00 in one hour. The conversion, then, is $\frac{8 \text{ dollars}}{1 \text{ hr}}$.

Question: If you earn \$8.00 per hour and work 78 hours in two weeks, how much will you earn?

$$78 \text{ hours} \times \frac{8 \text{ dollars}}{1 \text{ hour}} = \$624$$

If you want to have \$300 for a new TV set, how many hours will you have to work (without taking into account taxes)?

$$\$300 \times \frac{1 \text{ hour}}{\$8.00} = 37.5 \text{ hours}$$

Practice Problems *Answers on page 74*

1. Convert 4.20 feet to inches.
2. Convert 78 inches to feet.
3. Convert 1.9 meters to centimeters.
4. Convert 526 millimeters to meters.
5. Convert 6.09 kg to grams.
6. Convert 38 milliliters to liters.
7. Convert 0.5 kilometers to centimeters.
8. Convert 7.68×10^5 milligrams to kilograms.
9. Convert 2.50 inches to centimeters.
10. Convert 6.0 fluid ounces to milliliters

Study Aid 5 Conquering Tests

Preparing for the Test

- Test preparation begins on the first day of the course. Keeping current with all daily assignments is the most effective method of preparing for tests.
- Form a study group. You never really learn anything until you have to explain it to someone else. A study group will force you to do this. Working with others will also help you learn more study strategies. But study on your own both before and after the study group.
- Tests are more difficult than individual assignments because they contain a greater variety of material. To avoid getting confused, you must "overlearn" the material. To "overlearn," study actively until you really feel that you understand the material or can work the problems with no difficulty. Check to see how long you had to study or how many problems you had to work to reach this point. To "overlearn," you must spend at least half as much more time or work half as many more problems. Then you must also spend a small amount of time each day, maybe 10-15 minutes, reviewing past assignments after you have completed the new assignment.
- If there is a concept you do not understand, get help immediately. In chemistry learning is often sequential, i.e., new lessons build on past lessons. Therefore falling behind can be fatal!
- Try to predict test questions as you are studying.
- If practice tests are available, get them as soon as possible and do them at least several days before the test. Then you will still have time to get help if you find there are questions that are giving you difficulty.
- If you do fall behind in your studies, never try to learn new material the day or the night before a test. You will not have time to "overlearn" it. Therefore what will normally happen is you will confuse new material with what you already know. This confusion can cause you to miss even more test questions than you would have missed had you not tried to "cram in" the new material. Concentrate on what you know on the night before the test. Then, after the test is over, catch up immediately.
- The night before the test be sure to get enough rest. Also check your calculator and put it and any other materials you will need in a place where you will be sure to pick them up before leaving for class.

Taking the Test

- Wear a watch! Bring your calculator and make sure it is functioning.
- Get to class a few minutes early so you have time to get comfortable and get organized before the test is handed out.
- Put your name on the test!
- As soon as you get the test, write down on the test paper any information you are worried about forgetting. Then you won't have to worry anymore, and your mind will be free to concentrate completely on each test question.
- Preview the entire test. It only takes a minute or two but can tell you much.

  Previewing tells you:

  the number and types of questions the test contains (multiple choice, true-false, matching, essay, problems, etc.).

  the number of points each question is worth.

  any additional information you need to write down to be sure you remember it.

  how to schedule your time.
- Previewing should also build your confidence because, if you are prepared, you will know immediately that you can answer most of the questions.
- Read any directions carefully and completely. If there is something you do not understand, raise your hand and ask about it. The instructor will probably be able to clarify it for you. In any case, the worst that can happen is the instructor will say he or she cannot answer your question.
- Begin with the questions that are easiest for you. This strategy will build your confidence and lessen your anxiety. It may also remind you of things you need to know to answer the more difficult items. In addition, it will prevent you from losing points by spending too much time on things you don't know and running out of time before completing items you do know.
- Return to questions you skipped. If they are objective questions, i.e., true-false, multiple choice, matching, etc., make an "educated guess." (See *Mastering the Multiple Choice Test*). If they are essays or problems to be solved, do as much as you can. If the problem has several dependent parts and you do not know how to do the first part but do know how to do the next part, write a note saying, "I'm not sure how to do part A, so I am going to assume the answer is (make up a reasonable value) and use this value to complete the problem." Remember, part credit is better than no credit! Answer every question as completely as possible.

- Be neat and show your work, it will be easier to check your answers.
- Check your work! If you only have a few minutes left, check the first few questions you answered. You were most nervous then, and therefore you might have made careless errors. Then check the problems you thought were the easiest. You also make careless errors when you are too relaxed. If you eliminate careless errors, you can usually increase your grade by at least 10%!
- Do not get nervous because others are leaving before you. Those who finish early often are the weakest students. Use all the time allowed. There is no prize for finishing early!

Mastering the Multiple Choice Test

In addition to the general test-taking strategies we discussed at the end of Chapter 9, there are some specific things you can do to increase your score on multiple choice tests.

- As soon as you get the test paper, jot down any information you know you will need and are afraid you will forget.
- Scan the entire test. If any of the questions remind you of any other facts you are afraid of forgetting, write them down.
- Read the directions. Most multiple choice tests have only one correct choice, but don't assume anything.
- Answer the easiest, least time-consuming questions first. This strategy will assure that you have time to answer all the questions you know. It will also build your confidence and remind you of things that might help you answer the more difficult items.
- Before you look at the answer choices, work the problem and ask yourself "Does this answer make sense?" Remember that many of the incorrect answer choices will be derived from mistakes that students commonly make. If you see your answer as one of the choices, don't immediately mark it.
- If you don't find your answer among the choices, look again. Perhaps your answer is there but in a different form. For example 1/2 could be written as 0.5, 0.50 or 50%, etc. Maybe the answer is in scientific notation and you calculated your's in decimal form.
- After you've answered all the questions you know, carefully reread each question you skipped. You may find you now remember how to do them. Doing the questions you know warms you up and sometimes answering one question will spark your memory on how to answer another.

- If you are still unable to answer some of the questions, make an "educated" guess. Can you eliminate one or more of the choices? For example, if the question asks for grams, eliminate all the choices that don't have units of grams.
- If all else fails, simply guess. Choose a letter that is different from the question before and pick a letter that hasn't been used very often for a correct choice. Never leave a multiple choice question blank unless the directions say there is a penalty for guessing.
- Use all the time allowed. There is no prize for being the first to finish.

Solutions to Chapter 2 Practice Problems

1. 4

2. $8 - (-4) = 8 + 4 = 12$

2. –32

4. –2

5. –12

6. $-8 - (-4) = -8 + 4 = -4$

7. 32

8. 1/2 or 0.5

9. 1/6

10. $\frac{1}{2} \div \frac{1}{3} = \frac{1}{2} \times \frac{3}{1} = \frac{3}{2}$ or 1.5

11. $\frac{1}{2}$ of $8 = \frac{1}{2} \times 8 = 4$

12. 4 (11 and 12 are the same problem)

13. $\frac{1}{2} \div 8 = \frac{1}{2} \times \frac{1}{8} = \frac{1}{16}$

14. $8 \div \frac{1}{2} = 8 \times 2 = 16$

15. 0.5

16. 2.2

17. 0.3

18. 0.15

19. 0.13

20. 0.013

21. $\text{Percent} = \frac{\text{part}}{\text{whole}} \times 100 = \frac{40}{50} \times 100 = 80\%$

22. $\text{Part} = \frac{\%}{100} \times \text{Total} = \frac{90}{100} \times 60 = 54$ problems

23. 10^{13}

24. 10^{-3} or $\frac{1}{10^3}$ or $\frac{1}{1000}$

25. 10^{-3} or $\frac{1}{10^3}$ or $\frac{1}{1000}$

26. 10^{-1} or $\frac{1}{10}$

27. 80,000,000

28. 0.0013

29. 2.89×10^5

30. 5.4×10^{-5}

Solutions to Chapter 3 Practice Problems

1. 4.807 [×] 1.85 [÷] 0.467 [÷] 5 [=] **3.808543897**

   Rounded to 3 sig figs, the answer is 3.81.

2. 96.8 [÷] 15.2 [÷] 0.012 [=] **530.7017544**

   Rounded to 3 sig figs, the answer is 531.

3. 10.7 [×] 1.2 [÷] 3.8 [×] 14.2 [÷] 5.01[=] **9.577056413**

   Rounded to 2 sig figs, the answer is 9.6

4. 4.16 [EXP] 5 [×] 5.08 [EXP] 8 [+/-] [=] **0.0211328**

   Rounded to 3 sig figs, the answer is 0.0211. In scientific notation: 2.11×10^{-2}

5. 7 [EXP] 6 [+/-] [÷] 5.871 [EXP] 4 [=] **1.1923011** $^{-10}$

   Rounded to 3 sig figs, the answer is 1.19×10^{-10}

6. 5.68 [EXP] 3 [÷] 5.02 [EXP] 2 [=] **11.31474104**

   Rounded to 3 sig figs, the answer is 11.3. In scientific notation: 1.13×10^{1}

7. 1 [EXP] 5 [×] 1 [EXP] 8 [+/-] [×] 1 [EXP] 3 [=] **1**

   The answer is simply 1. Remember to enter the implied coefficients of 1.

8. 1 [EXP] 14 [+/-] [÷] 1 [EXP] 9 [+/-] [=] **0.00001**

   The answer in scientific notation is 1×10^{-5}

9. 3.15 [EXP] 18 [×] 342 [÷] 6.02 [EXP] 23 [=] **1.7895349** $^{-03}$

   Rounded to 3 sig figs, the answer is 1.79×10^{-3}. The decimal form is 0.00179

10. 1 [EXP] 15 [×] 1.66 [EXP] 24 [+/-] [÷] 12 [=] **1.38333** $^{-10}$

    Rounded to 3 sig figs, the answer is 1.38×10^{-10}

Solutions to Chapter 5 Problems

1. $4.2\,\text{feet} \times \dfrac{12\,\text{inches}}{1\,\text{foot}} = 50.4\,\text{inches}$

2. $78\,\text{inches} \times \dfrac{1\,\text{foot}}{12\,\text{inches}} = 6.5\,\text{feet}$

3. $1.9\,\text{m} \times \dfrac{100\,\text{cm}}{1\,\text{m}} = 190\,\text{cm}$

4. $526\,\text{mm} \times \dfrac{1\,\text{m}}{1000\,\text{mm}} = 0.526\,\text{m}$

5. $6.09\,\text{kg} \times \dfrac{1000\,\text{g}}{1\,\text{kg}} = 6090\,\text{g}$

6. $38.0\,\text{mL} \times \dfrac{1\,\text{L}}{1000\,\text{mL}} = 0.038\,\text{L}$

7. $0.5\,\text{km} \times \dfrac{1000\,\text{m}}{1\,\text{km}} \times \dfrac{100\,\text{cm}}{1\,\text{m}} = 50{,}000\,\text{cm} = 5\times10^{4}\,\text{cm}$

8. $7.68\times10^{5}\,\text{mg} \times \dfrac{1\,\text{g}}{1000\,\text{mg}} \times \dfrac{1\,\text{kg}}{1000\,\text{g}} = 0.768\,\text{kg}$

9. $2.50\,\text{inches} \times \dfrac{2.54\,\text{cm}}{1\,\text{inch}} = 6.35\,\text{cm}$

10. $6.0\,\text{fl. oz} \times \dfrac{1\,\text{qt}}{32\,\text{fl. oz}} \times \dfrac{1\,\text{L}}{1.06\,\text{qt}} \times \dfrac{1000\,\text{mL}}{1\,\text{L}} = 177\,\text{mL}$